FOR VIRTUAL REALITY AND MULTIMEDIA

FOR VIRTUAL REALITY
AND MULTIMEDIA

Durand R. Begault

AP PROFESSIONAL

Boston   San Diego   New York
London   Sydney   Tokyo   Toronto

About the Cover

Cylindrical Surface Plot of the Head-Related Transfer Function: Magnitude Response as a Function of Frequency over Azimuth Angle on a Radial Axis, by William Martens of E-mu/Creative Technology Center.

The cover photograph shows a visualization of the magnitude response (gain) of the head-related transfer function (HRTF) measured at the eardrum position of the anthropomorphic mannequin KEMAR. HRTFs were measured for 19 loudspeaker directions circling the side of the head facing the loudspeaker placed at ear level. The surface was constructed by interpolating the gain within each of 50 log-spaced frequency bands for the 19 HRTFs using a bicubic spline. The lowest band was centered on 55 Hz, the highest on 21,331 Hz. The distance of the surface from the origin and the color indicates the gain at a particular frequency and azimuth, which ranges from blue-black at the lowest gain (-43.9 dB) to a desaturated yellow at the peak gain (14.6 dB).

Page 13–14 quotation from: Gould, Glenn, "The Prospects of Recording," *High Fidelity,* April 1966. Used with the permission of Hachette Filipacchi Magazines, Inc., Estate of Glenn Gould, and Glenn Gould Limited.

AP PROFESSIONAL
1300 Boylston Street, Chestnut Hill, MA 02167

An Imprint of ACADEMIC PRESS, INC.
A Division of HARCOURT BRACE & COMPANY

United Kingdom Edition published by
ACADEMIC PRESS LIMITED
24–28 Oval Road, London NW1 7DX

Library of Congress Cataloging-in-Publication Data

Begault, Durand R., 1957–
    3-D sound for virtual reality and multimedia / Durand R. Begault.
      p. cm.
    Includes index.
    ISBN 0-12-084735-3 (acid-free paper)
    1. Human-computer interaction. 2. Virtual reality. 3. Multimedia systems. 4. Computer sound processing. I. Title.
    QA76.9.H85B44 1994   94-28168
    621.389'3--dc20   CIP

Printed in the United States of America
95 96 97 98 BB 9 8 7 6 5 4 3 2

Table of Contents

CHAPTER THREE
Overview of Spatial Hearing
Part II: Sound Source Distance and
Environmental Context   83

Preface

The process of designing a new human interface for a computing system — and, in particular, an improved audio interface for a multimedia or virtual reality system — can be a pursuit that is simultaneously mind-boggling and frustrating due to the complexity of the problem and the fact that humans, who inevitably mess up one's best-laid plans, are involved. The rewards are potentially very great, however: the synthesis and manipulation of auditory space embodies new domains of experience (or more properly, *tele-experience*) that promise to change the way we think about sound, how we manipulate it and experience it, and the manner in which the role of **listening** is evaluated in the future of computing.

The application of digital sound within computers, and in particular the manipulation of sound in **virtual space**, has only recently come to the attention of the general public. This contrasts the long-term development of tools for manipulating visual information in computing via graphical user interfaces (**GUIs**), video and still-image effects processing, and the sophistication of software available for graphics; after all, vision is (for most) the dominant perceptual sense and the principal means for acquiring information. So it is not surprising that computers have almost always had pretty good graphic and text manipulation resources; improvements in visual displays will continue to occupy both consumers and developers alike for a long time. Meanwhile, for various reasons, it seems to be a revelation to many that one can somehow improve on the annoying system **BEEP** that's emitted from a one-inch speaker within the computer.

Why has sound "played second fiddle" (in the fourth chair!) for so long? First, it isn't *absolutely* necessary for operating a computer; it's expendable, in the same way that one can drive an automobile (legally in most places) in spite of total deafness. Second, computer-based sonic manipulation was for a long time the almost exclusive domain of **computer musicians** and **psychoacousticians** — primarily professors and graduate students using the computing resources of universities or research and development facilities. Following the acceptance and standardization of digital sound in the recording industry and the diminishing price of digital technology in the 1980s, it only took a short time for digital musical instruments to become accessible to almost everyone. Consequently, the number of people with digital audio expertise or at least an appreciation of digital audio has mushroomed very quickly; it was inevitable that the notion of a more fully developed sonic interface for personal computers would become both desirable and consequently marketable. As an example, stereo digital sound is a consumer-expected, basic component of any off-the-shelf **multimedia system**. Although it hasn't happened yet to the same degree for **virtual reality**, an increase in the availability of cheaper, more accessible hardware will create more "experts" in the field with a need for expertise in implementing 3-D sound.

While this book mainly uses the phrase "3-D sound," equivalent designations include **virtual acoustics**, **binaural audio**, and **spatialized sound**. It's certain that as new products appear on the market, further variants on these phrases will appear. Fundamentally, all of these refer to techniques where the outer ears (the **pinnae**) are either directly implemented or modeled as digital filters. By filtering a digitized sound source with these filters, one can *potentially* place sounds anywhere in the virtual space about a headphone listener. The combination of 3-D sound within a human interface along with a system for managing acoustic input is termed a **3-D auditory display**. This is a highly general term that describes many kinds of sonic spatial manipulation and audio applications: integrated virtual reality systems, multimedia systems, communication systems, and entertainment. In other words, no particular concept, procedure, or theory can be assumed by a system that claims to include "3-D audio," beyond the inclusion of "pinnae filters."

A basic thesis of this book is that, first and foremost, users and developers of 3-D audio displays need to be aware of both the potential and the limitations of the current technology, since the excitement surrounding a potential application can quickly be diminished if expectations are unreal or if planning and design are incomplete. The best way to ensure success and to envision better systems for the future is to provide an understanding of how important issues in psychoacoustics and engineering interact. For this reason, this book (in particular, Chapters 2 and 3) emphasizes the **psychoacoustics** of spatial hearing: the relationship between the objective physical characteristics of sound and their subjective interpretation by listeners. This type of approach prioritizes the "human" in the interface over the "buzzers and bells" of the technology.

The ability to meld basic research with an applications vision is a difficult task if only for the reason that the necessary expertise and information is spread among many disciplines and resources. As author Alvin Toffler (*Future Shock*) has suggested, society has evolved to a point where acquisition of information is no longer a problem; rather, the challenge is how to navigate through it, since much of it is useless for our own purposes (as any reader of news groups on the Internet knows). The design of many aspects of "modern" human computer interfaces (e.g., mice, graphic interfaces, and wrist rests to prevent carpal tunnel damage) has only evolved as a result of the interaction of many types of expertise from a diverse collection of fields.

The application of 3-D sound within multimedia or virtual reality is no different. Consider how strange the diversity of a good design team might be to the uninitiated. The **computer programmer**—that almost holy personage who lies mythically somewhere between Mr. Goodwrench and our presupposed vision of the hypothetical "nerd"—is suddenly as indispensable as the **psychophysicist**, a profession ridiculed in the first minutes of the popular movie "Ghostbusters" as a scientist walking the line of charlatanism (sure, they know about the perception of clicks and sine waves, but what about sounds that exist anywhere but in a laboratory?). A close relative of the psychophysicist is the **human factors engineer**, who at best must contribute to new designs for orthopedic shoe inserts. Then there's the **electrical engineer**—a near relative of the computer programmer who for some reason finds the mathematical pole-zero plots of his or her filters ultimately more self-satisfying than the "for loop" that make a programmer delight because his or her code contains 13 fewer lines. Let us not forget the **hardware designer**, who deals with assembly code and Ohm's law on a level that can seem to the uninitiated at once archaic, magical, and impenetrable! And finally, there's the **audiophile**, whose expertise needs to be tapped at some point.

No one is perfectly characterized by these stereotypes, and most people who are interested in 3-D sound have varied combinations of expertise. But from a design and systems engineering standpoint, all of these areas are critical for eventual successful applications of 3-D sound. It's assumed then that to integrate sound into multimedia and virtual interfaces, a global view of knowledge pertaining to the problem would be beneficial. In other words, what's the big picture as far as getting computer users excited about sound as a new way of interfacing with machines? How can this interface be best facilitated? What are the important issues for successful implementation?

While many are excited by the potential of spatial sound manipulation—from composers to multi-media display designers, from recording engineers to aeronautic flight deck designers—there are few complete sources on the topic as a whole, and none that are simultaneously introductory and comprehensive from an applications standpoint. In this work, the reader will find a guided tour through the physics and psychoacoustics of sound and spatial hearing, the means for implementing 3-D sound, and diverse examples of how virtual acoustics are applied.

Overview

Chapter 1, "Virtual Auditory Space: Context, Acoustics, and Psychoacoustics," describes the basic components of virtual reality and multimedia systems, how their configurations differ, and how 3-D sound fits in, from the perspectives of both the listener and the system designer. Terminology useful for describing virtual audio experiences is introduced, concluding with a brief overview of the basics of the physics and psychoacoustics of sound relevant to understanding spatial hearing.

Chapter 2, "Overview of Spatial Hearing. Part I: Azimuth and Elevation Perception," reviews the physical features and psychoacoustics related to spatial hearing: how sounds are heard below–above, front–behind, and left–right. Summaries of important localization studies are brought together in a manner useful for implementing and assessing the application requirements of a 3-D audio system.

Chapter 3, "Overview of Spatial Hearing. Part II: Sound Source Distance and Environmental Context," focuses on how judgments of sound source distance and the space surrounding the sound source are made. The basics of the physical and psychoacoustic features of reverberation are reviewed. As in Chapter 2, relevant psychoacoustic data are summarized with a viewpoint toward implementation.

Chapter 4, "Implementing 3-D Sound: Systems, Sources, and Signal Processing," shows the basics of digital signal processing that allow modeling spatial hearing cues. Included are a review of the design of useful sounds for input to a 3-D sound system, the means for obtaining and synthesizing spatial cues, and an introduction to digital filters.

Chapter 5, "Virtual Acoustic Applications," reviews a variety of virtual acoustic displays that are as varied in context as they are in the sophistication of the implementation. Information on spatial sound interfaces for entertainment, computing, aeronautics, and communications is reviewed. This includes warning systems for high-stress human machine interfaces, speech intelligibility for radio communications, sonic windowing systems, and auralization— the ability to simulate and listen to a sound within a virtual room.

Chapter 6, "Resources," concludes the book. For those who want to get started in 3-D sound or for the person who has merely lost an important address, this Chapter offers the names and addresses of most of the important software and hardware manufacturers relevant to virtual acoustics. For those who desire additional information on the topics in this book, an annotated listing of journals, books, proceedings and patents is given.

Acknowledgments

Without the assistance and support of Justine, Pat, Denise, Chloe, Nigel, tre, Bill, Lars, Joel, Rick, Dr. B, Jenifer, Reuben, and many others, the completion of this book would not have been possible.

CHAPTER ONE

Virtual Auditory Space

Context, Acoustics, and Psychoacoustics

CONTEXT

How can we verbally characterize the auditory spatial percepts of a listener? More importantly, which of these percepts might be controllable with a virtual acoustic display, i.e., a **3-D sound system**?

The idealistic but laudable goal of a 3-D sound system designer ought to involve the notion of complete control and manipulation of someone else's spatial auditory perception. In other words, by manipulating a machine—either in real time or algorithmically—the operator of a 3-D sound system will want to predict, within a certain level of allowable statistical predictability, the auditory spatial imagery of a listener using the 3-D sound system. This is referred to as **spatial manipulation.** In some cases, the designer of the 3-D sound system is the operator—and sometimes the operator is the listener—but often, predictability and control over the spatial perception of any listener are desirable. It follows that control over virtual acoustic imagery involves not only engineering based on physical parameters, but also psychoacoustic considerations.

A 3-D sound system uses processes that either complement or replace spatial attributes that existed originally in association with a given sound source. Only imagination and the tools at hand pose limits to the spatial manipulations possible for a user to make with existing technology. Malleable virtual acoustic simulations, such as those described in the following examples, could be produced for a listener:

1. **Replication** of an existing, spatial auditory condition. Starting with a recording of a concert violinist made in a relatively reverberant-free environment, one could, with absolute control, create a virtual acoustic experience that gives rise to a percept of the violinist walking back and forth on the stage of a well-known auditorium like Carnegie Hall, from the perspective of any seat in the auditorium.

2. **Creation** of completely unknown spatial auditory experiences. For instance, shrinking the violinist to a sound source the size of a fly and then hearing it play while it zooms rapidly about your head in a marble bathroom the size of a football stadium.

3. **Transmutation** between one spatial experience and another. You're listening as if seated on the stage of Carnegie Hall. The violinist, playing a furious solo *cadenza*, moves rapidly toward you across a quiet, snow-covered field from 400 feet away. The music gets closer and closer, wider, more present, and more reverberant until, after a minute, right at the climax of the solo, the violinist arrives at the front of the stage, joined by the orchestra.

4. **Representation** of a virtual auditory world. Through headphones, a "slice of life" of the acoustic experience of a violinist could be experienced, as heard through the violinist's ears. The user of the 3-D sound system would use an interface to be transported at will from the limousine to the green room, then backstage behind the curtain before the concert, and then out a side stage exit back to the street. The 3-D sound system designer has in this case acoustically mapped the acoustic environment in advance, as well as the means for allowing the listener to control his or her interaction in the environment.

Figure 1.1 shows a taxonomy of spatial perception attributes that are referred to throughout this book. Fundamentally, spatial perception involves an egocentric frame of reference; measurements and orientation of sound images are given from the **listener**'s position. In the psychophysical studies reviewed ahead, the reference point for describing distances and angles of sound images is located at an origin point directly between the ears, approximately at eye level in the center of the head. In addition, there is the **virtual sound source** (sound image) to be localized; and a linear distance between it and the listener to be interpreted as **perceived distance.**

A special case is where the listener is the sound source; in this case, there may be no perceived distance between one's self and one's own voice, although in some applications, such as radio communications, one's own voice is heard through a headset as **sidetone**, usually at one ear.

Figure 1.1 also shows two conventions for describing the angular perception of a virtual sound source: **azimuth** and **elevation**. Azimuth perception is particularly robust, since human ears are located at almost opposite positions on either side of the head, favoring hearing of the relative angle of sound sources on a plane parallel to the surface of the ground. Normally, azimuth is described in terms of degrees, where 0 degrees elevation and azimuth are at a point directly ahead of the listener, along a line bisecting the head outward from the origin point. In some systems, azimuth is described as increasing counterclockwise from 0–360 degrees along the azimuthal circle. For descriptive purposes it is often more convenient to describe leftward movement along the azimuth, between 0 and 180 degrees, as **left** 0–180 degrees, increasing counterclockwise; and positions to the right of the listener as **right** 0–180 degrees, increasing clockwise. Elevation increases upward from 0 to a point directly above a listener at **up** 90 degrees, or directly below at **down** 90 degrees. This polar-based system is used throughout this book. In actual software implementations, it is better to use a 0–360 degree system, or **Euler angles** (given in terms of **pitch**, **yaw**, and **roll**). The perceptual issues related to azimuth and elevation will be reviewed in detail in Chapter 2.

Azimuth and elevation descriptions can indicate a sound source's perceived position only in terms of its location on the surface of a sphere surrounding a listener's head. A more complete description would include the perceived **distance** of the sound source as another dimensional attribute (Figure 1.1). Amazingly, our awareness of distance is quite active, yet often inaccurate.

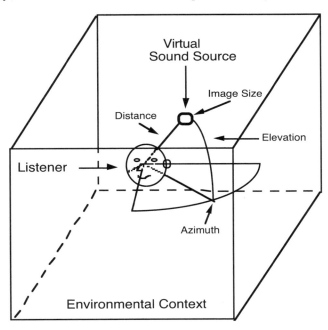

FIGURE 1.1. A taxonomy of spatial manipulation (from the operator's perspective), or of spatial hearing (from the listener's perspective).

This is because distance perception is **multidimensional**, involving the interaction of several cues that can even be contradictory. One can speak of "relative distance," the proportional increase or decrease in distance of a virtual sound source from an arbitrary location, or of a specific "absolute distance" using feet or meters as a reference. For example, "the bass drum sounds twice as close as the singer, and the piano sounds somewhere in-between" are relative distance estimates that are useful for most purposes. On the other hand, "the buzz saw moves to an inch from your head" is an absolute estimate of distance. Distance perception is reviewed in Chapter 3.

When we describe the location of a sound source, it is natural to think of the center of the source as a point that describes its actual location. For instance, taking cities within California as an example, Los Angeles sprawls over a much larger area than San Francisco, but we describe the location of both cities in terms of a point location, typically in terms of degrees latitude and longitude (usually to the center of City Hall). Like cities, the perceived image of the sound source will have a particular **width** or **extent** as well, as shown in Figure 1.1 by the size of the circle representing the sound source. Everyday listening suggests that sounds seem to occupy larger or smaller areas of acoustic extent, across azimuth and elevation. Like distance, the cues for this particular spatial percept are also multidimensional; **auditory spaciousness** and **apparent source width** are reviewed in Chapter 3.

Finally, there is the **environmental context**. This primarily refers to the effects of **reverberation**, caused by repeated reflections of a sound source from the surfaces of an enclosure (see Figure 1.2). Like a light source on a painting, the sound reflections from the surfaces of an enclosure or in the outdoor environment can potentially cause a significant effect on how a sound source is

FIGURE 1.2. A simple, two-dimensional model of how a listener (L) in an enclosure will hear both direct and indirect sound paths from a sound source (S). The distribution of the indirect paths—reverberation—yields information for perception both about the sound source S and the environmental context E.

perceived. The effect of the environmental context is manifested by a set of secondary sound sources dependent on and made active by the primary localized sound source. There are two categories of perceptual effects caused by this: (1) an effect on a given virtual sound source's location, and (2) the formation by the listener of an image of the space occupied by the sound source. The latter is articulated by the sound source, but cannot be a source itself.

In order to evaluate spatial hearing from either a psychoacoustic or engineering standpoint, it is necessary to consider all significant factors involved in the sound transmission path. So far, the focus has been on the potential control of the operator; but realistically, the nature of the 3-D sound system and the degree of control available will be a function of the particular application, and the degree to which human perception is taken into its design. It is most important to ascertain the realistic limitations for establishing a match between the listener's and the operator's spatial hearing experience.

Ahead, a sequential overview is made of some of the important events that occur in the transformation of a sound source into a spatial percept within a listener's mind. First, a contextual description of the disposition of 3-D sound within virtual reality and multimedia systems is given. Second, a brief overview of descriptive terminology for describing sound sources relevant to later chapters is given. Third, some general perceptual issues are covered, including cognition. Chapters 2 and 3 cover in greater detail the psychoacoustic mechanisms for spatial interpretation of sound source location, particularly its angle and elevation relative to the listener, distance perception, and the effects of the environmental context. Chapter 4 examines the implementation of 3-D sound, based on the perceptual information presented; Chapter 5 shows example applications; and Chapter 6 lists resources for additional information.

Source-Medium-Receiver Model:
Natural versus Virtual Spatial Hearing

A distinction must be made between experiences heard in most everyday activities, and those experiences heard over headphones using an audio reproduction system. **Natural spatial hearing** refers to how we hear sounds spatially in everyday hearing, with our ears uncovered, our head moving, and in interaction with other sensory input. Note that spatial auditory images are not confined to "two-ear," or "binaural" hearing; one-ear hearing can provide spatial cues, as proven by experience with telephone conversations. A special case of binaural hearing is **virtual spatial hearing;** this refers here to the formation of synthetic spatial acoustic imagery using a **3-D sound system** and stereo headphones. Virtual or natural spatial hearing situations can both give rise to the percepts described in Figure 1.1.

The transmission path from operator to listener can be described according to simple, or increasingly complex, communications theory models. These models

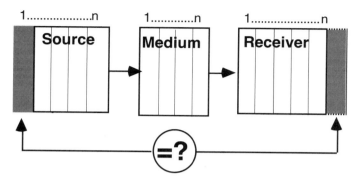

FIGURE 1.3. A source-medium-receiver model. Each element contains a number of physical, neurological, or perceptual transformations (dotted line sections numbered 1 through n) in the communication of the spatial location of a sound source. For a 3-D sound system, the question is to what degree the first element of the source—the specified position indicated by the operator—equals the nth element of the receiver, the ultimate spatial perception of a virtual acoustic image.

at root involve a **source**, **medium**, and **receiver** (see Figure 1.3). For a psychoacoustic experiment, the purpose of such a model is to illustrate how manipulation of a particular independent variable of sound will give rise to changes in spatial imagery. For a 3-D sound system, the purpose is to elucidate the difference between the operator's specification of a particular spatial image and the listener's spatial imagery, as a result of the requirements of the particular application.

While the *source* can involve one or more vibratory sources, it is most convenient to describe the system in terms of a single source in isolation. Natural spatial hearing seldom involves a single source, whereas virtual spatial hearing with a 3-D sound system involves the positioning of each of a number of sources individually. The *medium* involves the paths by which the source arrives at the listener. In natural spatial hearing, this involves the environmental context (reverberation and the effects of physical objects on the propagation of the sound); in a 3-D sound system, sound reproduction nonlinearities, signal processing, and headphones become the major components. Finally, the *receiver* involves the listener's physiology: the hearing system, from the ear to the final stages of perception by the brain.

Ultimately, it is the **nonlinear transformations**, or **distortions** that occur at various "significant way points" along the source-medium-receiver chain (each of the sections 1–*n* in Figure 1.3) that are useful to identify in both natural and synthetic spatial hearing contexts. The nonlinear transformations within a source-medium-receiver chain are analogous to the party game "telephone," where a verbal message is whispered secretly in a sequential manner between several persons—the game's amusement value is in the difference between the original message and its result at the last person. From an engineering standpoint, a perceptually meaningful, segmented approach for effecting synthetic binaural hearing can result from understanding natural spatial hearing. The nonlinearities between the various source-medium-receiver stages can be described statistically via psychoacoustic research results related to 3-D sound. Such research attempts to describe the degree of mismatch between the operator's intended spatial manipulation and the listener's perception, based on the probability of responses from the overall population for a limited set of independent variables.

First, take the case of natural spatial hearing. Figure 1.4 depicts a schematic transformation between sound source and listener through a medium. The sound source consists of undulations within the elastic medium of air resulting from the vibration of a physical object, such as vocal cords within a larynx, or of a loudspeaker cone. If the sound source propagates in a manner irrespective of direction, it can be said to be **omnidirectional**, and it therefore forms a **spherical field** of emission. If the sound source is within an environmental context without reflections—e.g., an **anechoic chamber**—then beyond a certain distance the sound waves arrive at the front of a listener as a **plane field**, meaning that the sound pressure would be constant in any plane perpendicular to the direction of propagation. But within nonanechoic environmental contexts, sound arrives to a listener by both direct and indirect paths, as shown in Figure 1.2. In this case, the sound source can be said to arrive at the listener as a **diffuse field**, due to the effect of the **environmental context**. The environmental context is therefore the

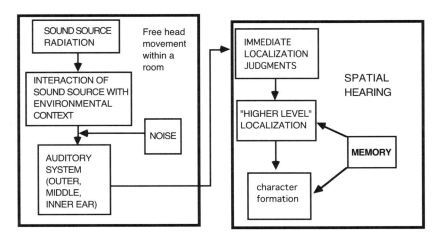

FIGURE 1.4. A model of natural spatial hearing.

principal component of the *medium* in the case of natural spatial hearing. Other sound sources within the environmental context might be considered undesirable; these are generally classified as **noise** sources. The manipulation and diminution of the noise experienced by most people is often within the domain of architectural acoustic and community noise-control specialists; sometimes their solutions are inconspicuous, as in a standard for noise isolation for a building wall, and other times conspicuous, as with freeway or airport sound barriers.

Next, consider the source-medium-receiver communication chain for a hypothetical 3-D sound system (see Figure 1.5). Assume that two-channel (stereo) headphones are used to reproduce a sound source. In this case, an actual environmental context is absent, and it must be simulated in most cases. The source must either be **transduced** from acoustic to electrical energy via a microphone, obtained from a storage medium such as a CD or a disc drive, or produced by a **tone generator**—a generic phrase for samplers, synthesizers, and the like. In addition, the receiver is no longer moving his or her head in the same environment as the sound source. Instead, a pair of transducers, in the form of the headphones placed against the outer ears, become the actual source and medium. This type of interaction will of course affect an incoming sound field differently than with natural spatial hearing. In addition, if the sound source should stay in a single location in virtual space, then its location relative to the listener's head position will also need to be synthesized and updated in real-time.

The "black box" that comprises the arbitrary spatial sound processor will transform the electrical representation of sound in some way, activating the spatial hearing mechanism's perceptual and cognitive aspects to work together to form a particular spatial judgment. This spatial judgment could be one similar to natural spatial hearing as shown in Figure 1.1, or could be a spatial hearing experience only possible with the given apparatus, and not heard within natural spatial hearing. This contrast underlies an important difference between the phrases **virtual environment** and **virtual reality**—no correlation to reality is necessarily assumed grammatically by the first term, and we may certainly want to create virtual experiences that have no counterpart in reality. In

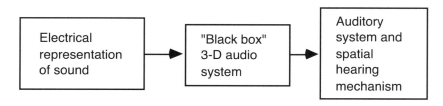

FIGURE 1.5. A model of virtual spatial hearing.

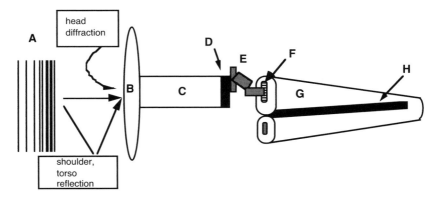

FIGURE 1.6. Highly simplified, schematic overview of the auditory system. See text for explanation of letters. The cochlea (**G**) is "unrolled" from its usual snail shell–like shape.

fact, the phrase "naive realism" has been used to critique virtual environment designers that purport to effect perceptual experiences that are as real as those experienced under natural conditions.

At the listener, natural and synthetic spatial hearing experiences consist of both physical and perceptual transformations of an incoming sound field. For the moment, the basic physical system of the ear is reviewed through the following thumbnail sketch of the auditory system (see Figure 1.6). Sound (**A**) is first transformed by the **pinnae** (the visible portion of the outer ear) (**B**) and proximate parts of the body such as the shoulder and head. Following this are the effects of the **meatus** (or "ear canal") (**C**) that leads to the middle ear, which consists of the **eardrum** (**D**) and the **ossicles** (the small bones popularly termed the "hammer-anvil-stirrup") (**E**). Sound is transformed at the middle ear from acoustical energy at the eardrum to mechanical energy at the ossicles; the ossicles convert the mechanical energy into fluid pressure within the inner ear (the **cochlea**) (**G**) via motion at the **oval window** (**F**). The fluid pressure causes frequency dependent vibration patterns of the **basilar membrane** (**H**) within the inner ear; which causes numerous fibers protruding from auditory **hair cells** to bend. These in turn activate electrical action potentials within the neurons of the auditory system, which are combined at higher levels with information from the opposite ear (explained more in detail in Chapter 2). These neurological processes are eventually transformed into aural perception and cognition, including the perception of spatial attributes of a sound resulting from both monaural and binaural listening.

APPLICATION TYPES

Within the applications of virtual reality and multimedia lies a wealth of confusing or even contradictory terminology that is inherent to any young technology. **Virtual reality systems** would be better described as **virtual environment systems**, since reality is often difficult to simulate, and it's usually better to simulate something other than "reality." **Multimedia systems** on the other hand would be better described specifically as what *kinds of media* are involved in the particular application. Both fields are similar in that they are at root *multiperceptual, real-time interactive communication media.*

Like all "buzzword" concepts that engender great potential financial and scientific value, virtual reality and multimedia are represented by a seemingly endless number of forms, applications, and implementations. How one actually *defines* either area or distinguishes between them is in itself an endless source of conferences, articles, and speculation. As time progresses, the tools and concepts of virtual reality and multimedia increasingly overlap. Pimentel and Teixeira (1992) distinguish between multimedia and virtual reality by using a description given by author Sandra Morris: "...virtual reality involves the creation of something new, while multimedia is about bringing old media forms together to the computer." The distinction is fairly useless, though, if for no other reason than "old" media quickly results from "new" media. Hardware from virtual reality can be integrated into a multimedia system, and virtual reality often involves older media forms.

Virtual reality and multimedia systems can be described in terms of the diagrammatic representation shown in Figure 1.7. A hardware component definition of a virtual reality system usually involves the following elements:

1. One or more host computers for scenario management, termed a **reality engine** by Teixeira and Pimentel, that contain algorithms useful for organizing and creating visual, auditory, and haptic illusions for one or more users according to a particular scenario (popularly termed a **virtual world**).
2. **Effectors**, hardware devices such as helmet-mounted displays, headphones and force-feedback devices that cause the illusion at the user to occur in response to data contained within the reality engine.
3. **Sensors**, hardware such as magnetic or mechanical position trackers, six-degrees-of-freedom trackballs, speech analyzers, and data gloves.

The reality engine can process information to and from multiple effectors and sensors for one or several persons or machines. When controlling a machine such as a robot, tool, or vehicle-mounted camera, the word **telerobotics** is used.

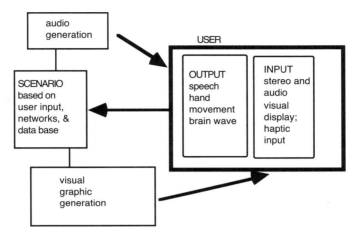

FIGURE 1.7. Components of a virtual reality or multimedia system. The scenario software is termed a "reality engine" within virtual reality; in a multimedia system, it is the software program that includes "authoring software." The user receives **effector** data, primarily in the form of visual and audio input, and interacts with the scenario by outputting data from various **sensors**.

In contrast to a virtual reality's system of effectors, sensors, and reality engine database, a **multimedia system** can be defined simply as any actual or near real-time interactive software interface that includes multiple effectors, including the addition of "high-quality" sound (i.e., something more sophisticated than system "beeps"). The sound component is easily available on plug-in **sound cards**, such as Creative Labs' SoundBlaster®. The multimedia system can also include advanced graphics, animation, and/or video-based visual interfaces, and possibly advanced types of sensors such as speech recognition. Desktop computer systems such as the Apple *Macintosh*® Quadra series or the Silicon Graphics *Indy*® combine high-speed graphic chips for interactive rendering of screen images and simple but useful software for including "CD-quality" sound with images; equivalent systems can also be assembled "from scratch" by adding sound and graphics accelerator cards to an existing personal computer.

The introduction to the consumer market of the various CD formats (such as **CD-I** (compact disc interactive) and **CD-ROM** (compact disc read-only memory) has enabled faster access to large databases, enabling higher-quality rendering of pre-stored graphic images and audio (see Pohlmann, 1992). The result has been the creation of many new types of desktop-based interactive media. The user is allowed a greater level of involvement in a narrative presentation than with normal software in that participation is solicited by the software to determine its final form. This can be as simple as telling the program that you want to see the next screen by pushing a button, or as complicated as being involved as the **author** of a multimedia presentation. The latter is accomplished using **authoring software**, which can include **toolboxes** useful for organizing a multimedia presentation. In other words, a nonprogrammer can script a presentation without writing actual computer code. However, at root the user is constrained by the available authoring systems; some professionally produced interactive media have cost millions to develop by multimedia software laboratories. It is especially true that the requirement for the successful future of both virtual reality and multimedia will be for software to match the potential of the hardware.

One realization that quickly becomes apparent for those first venturing into virtual audio and multimedia is that the development of sound within human interfaces has not been developed to anywhere near the level that visual interfaces have. For instance, high-resolution color graphic hardware and software have been around longer on personal computers than the audio equivalent, CD-quality two-channel digital sound.

These component-based definitions only characterize the hardware of the art. What about the detail or quality of the information contained within the reality engine portion of the system? Or the match between the particular implementation of the system to a real-world application? The range of components varies widely in systems promising virtual reality, extending from the familiar VPL-type hardware with interfaces such as helmets, gloves, and bodysuits, with high-quality visual rendering—to a game played on a Macintosh at one's desk by interacting with a standard mouse. Multimedia on the other hand can mean anything from a desktop virtual reality system to a cheesy screen animation of a corporate presentation, combined with sound taken from the "Rocky" soundtrack.

In the passive entertainment realms such as video, recorded music, or movies, a one-way communication medium is used instead of the two-way communication media featured in multimedia and virtual reality. However, while the experience of watching a movie in the theater is noninteractive, it can *at times* seem more realistic than a virtual reality experience. This is because the bulk of the technology in a passive medium is based on presenting visual and audio images from a single perspective. Virtual reality systems put the bulk of their technologies in passing data between components to allow real-time or near real-time interactivity, maximizing the limits of computer resources to perform the analysis of the sensors and production of high-quality rendering of pictorial and auditory environments at the effectors. However, it is both the interactivity and the quality of the rendering that results in the *immersiveness* of a virtual reality or multimedia system.

Like many things, virtual reality and multimedia systems are highly varied in design and purpose. The degree to which the application can be personalized functions to increase the level of immersiveness of the interface and optimizes the interaction between the individual user and the specific application. This is probably largely a cultural phenomenon dependent on current expectation, meaning that yesterday's immersiveness is today's motivation for purchasing an improved product. If, on the other hand, laboratory standards for determining immersiveness for the general population were known, they would probably be written into law. We might then be able to obtain the virtual immersiveness percentage contained within a system by reading a special label on the side. (If one were caught operating heavy machinery while immersed in a virtual reality apparatus rated 75% or above immersion, there could be an argument that public safety laws were in violation—off to the city cyber tank for 48 hours.)

If the quality of *immersiveness* seems a bit nebulous, it can at least be viewed as a general criterion for dictating specific requirements for perceptual *performance* and especially *simulation*. For performance, the ease of use and productivity resulting from using a virtual reality or multimedia system as a communication medium might be tied to the quality of immersiveness. Perceptual simulation criteria can only be established in light of human-factors–based psychophysical investigations, where performance and perception can be measured statistically. Often, both performance and simulation are affected as a function of the cost of the hardware placed into the system, which is tied directly to the quality of the optics, the speed of the computers, and the reliability of the sensors.

Occasionally, one can see in the visual world a solution to a quality problem by compromising another factor, such as interactivity. The *BOOM*™ (Binocular Omni-Orientation Monitor) manufactured by Fakespace Incorporated (Menlo Park, California) requires one to peer into a counterweighted, yoke-mounted display in order to see a virtual world, but has an improved visual display because it can use cathode ray tube (CRT) hardware instead of the lighter but less robust liquid crystal displays (LCDs) found in helmet-mounted displays (Bryson and Gerald-Yamasaki, 1992). But another solution for improving the immersivity and perceived quality of a visual display and the virtual simulation in general is to focus on other perceptual senses—in particular, sound. A New York Times interviewer, writing on a simulation of a waterfall by a group at the Banff Center for the Arts headed by virtual reality developer Brenda Laurel, described how "the blurry white sheet" that was meant to simulate a waterfall through a $30,000 helmet-mounted display seemed more real and convincing with the addition of the spatialized sound of the water (Tierney, 1993). Dr. Laurel went on to comment how, while in the video game business, she found that "really high-quality audio will actually make people tell you that games have better pictures, but really good pictures will not make audio sound better; in fact, they make audio sound worse. So in the (virtual) world we're building, one of the features is a rich auditory environment."

Virtual Audio: A Special Case

Compared to the synthesis of virtual visual experiences, the transformation of experience via acoustic storage and simulation of spatial audio is not only much easier to accomplish, but is also usually more convincing. For example, consider the popular genre of environmental-nature recordings that are binaurally recorded expressly for the purpose of "transforming one's moods to a gentle northeastern farmland at dusk," etc. There is a rich history of radio drama production that attests to the power of combining monaurally delivered sound for narrative, dramatic purposes (e.g., Orson Wells' "War of the Worlds"). Intuitively it seems that audio is simply *easier* to use for transportation into a virtual world, especially if we keep our eyes closed and free from distractions.

Real-time interactivity is also not a necessary requirement; for instance, 3-D audio might be considered by some to be *virtual* in the way it can mimic a spatial audio experience. From that perspective, it may consist of nothing more than a cassette that is available from a soft drink can. Specifically, the summer of 1993 saw one of the more commercial displays of 3-D technology since its use on a Michael Jackson album—cans of Hawaiian Punch® with the following offer: "TAKE A FANTASTIC JOURNEY IN 3-D SOUND. JOIN PUNCHY® ON AN INCREDIBLE 3-D SONIC ADVENTURE. FIND YOURSELF TRAPPED INSIDE THE CHANNELS ON A TELEVISION WITH ONLY MINUTES TO ESCAPE!" Here, the imagination, triggered by the literary suggestion, is the source of the user's interactivity.

If we segregate audition from other percepts, we find that artificial sound environments are part of our everyday lives. Virtual auditory environments are familiar to everyone through daily exposure to loudspeakers, to the extent that many persons' experience of music is more familiar via prerecorded media than live performance. For most composers, the notion of what an "orchestra" sounds like is the result of the recording media. Pianist Glenn Gould characterized the function of musical recordings in society as follows:

> If we were to take an inventory of those musical predilections most characteristic of our generation, we would discover that almost every item on such a list could be attributed to the influence of the recording. . . . today's listeners have come to associate musical performance with sounds possessed of characteristics . . . such as analytic clarity, immediacy, and indeed almost tactile proximity. Within the last few decades, the performance of music has ceased to be an occasion, requiring an excuse and a tuxedo, and accorded, when encountered, an almost religious devotion; music has become a pervasive influence in our lives. . . . The more intimate terms of our experience with recordings have since suggested to us an

> acoustic with a direct and impartial presence, one with which we can live in our homes on rather casual terms. (Gould, 1966)

Even the two-inch speaker in your television is capable of producing a non-interactive, virtual auditory environment; without looking at the screen, we can easily imagine the relationship of a live audience to a talk show host and her guests, an intimate dialogue or confrontation between two people, or the completely synthetic experience of a "talking head" such as a newscaster. The virtual auditory world of theater and opera was taken into radio and television dramas; hence, it is perfectly normal today in the virtual auditory world of TV to hear a rock band wailing away in the same virtual auditory space of police cars screeching around corners in hot pursuit, or string orchestras in the same room as the unfaithful lover giving a confessional.

We've all been molded to and even welcome in a subconscious way the opportunity to have a willing suspension of disbelief regarding the sound combinations we hear. Of course, it's our imagination, or more properly, our higher-level cognitive associations with previous memory of how these situations look and sound, that allows us to accept the vibrations of a two-inch transducer as a form of virtual acoustic reality.

Our perceptual system is often explained as being dominated by vision. Indeed, although a loudspeaker may be displaced from the actual location of a visual image on a television or movie screen, we can easily imagine the sound as coming from an actor's mouth or from a passing car. This is an example of what is termed **visual capture**; the location of the visual image "captures" the location derived from audio cues. The power of visual capture is exemplified by the fact that film prints made for distribution usually guarantee no less than a 40-msec delay between optical audio tracks and the visual image.

The relationship between television sound and virtual audio imagery suggests that, in spite of modest technology such as a small loudspeaker, cognitive associations are very powerful for modifying how sound is interpreted, and even where the sound is interpreted to be in space. The point is that a sound in itself has considerable "virtual potential." For instance, it is currently expensive to provide tactile feedback in a virtual environment, but it is possible to use sound to suggest tactile experiences. An example is the problem of simulating the tactile sensation of walking into a virtual wall; hence, some have provided a "crashing" sound through headphones when a person walks into a virtual boundary. Similar sounds were used as part of the Matsushita VPL-based virtual reality system, designed to immerse potential customers in the world of custom-designed kitchen spaces. The use of specific sounds for **auditory feedback** is discussed further in Chapter 5.

Components

Early on in the development of virtual reality, it was seen that sound should have some type of interactive, spatial character imposed on it. At NASA Ames Research Center in Mountain View, California, the development of one of the first "goggles and gloves" virtual reality systems in the mid-1980s by Scott Fisher and his colleagues was paralleled with the development of one of the first interactive, virtual auditory displays (Wenzel, Wightman, and Foster, 1988). The 3-D audio display termed the **Convolvotron**™ by its developers, Elizabeth Wenzel and Scott Foster, was developed separately from the visual portion of the NASA virtual reality system, and then integrated at a later point. This is typical of the development of audio within a virtual reality system design effort; as a consequence, it often exists in hardware and software as a separate system from the host "reality engine" computer.

Figure 1.8 shows a general configuration for the interface of a sound subsystem in a virtual reality system. The host computer calls on databases from the reality engine that describe the acoustic world, along with input from sensors such as trackers that allow updating the position of the virtual sound source in relation to the position of the head. The data and sensor input are combined into a set of multiplexed instructions that are sent to the audio subsystem. The audio subsystem contains two fundamental parts: a **signal**

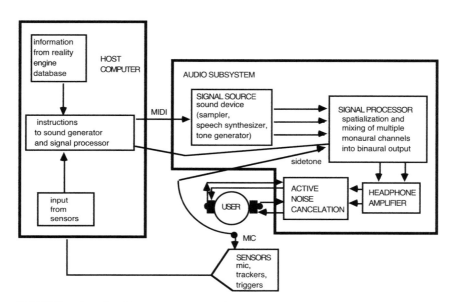

FIGURE 1.8. Configuration of audio components, effectors and sensors within a virtual reality system.

source, such as a synthesizer, and a **signal processor** that contains components for spatializing and mixing the outputs from the signal sources. The audio subsystem might also include **active noise cancellation** technology. These components are explained in detail in Chapters 4 and 5.

Most multimedia systems do not include 3-D sound (yet!). Unlike many virtual reality systems, sound storage and processing are contained entirely within the host computer. Often, the sound is prerecorded and then subjected to compression algorithms, which can degrade quality considerably from that found on a regular CD, where audio is *not* compressed. "Multimedia upgrade kits" introduced to the general consumer market include a **sound card**, which contains the necessary components for recording sound and then playing it back, and software for arranging and editing the sound in relation to visual images. These cards include miniature **sound synthesizers** (also called **tone generators**), usually based on either **wave tables** (sampled sound) or **frequency modulation (FM) synthesis.** Presently, only the most advanced multimedia systems contain hardware capable of **digital signal processing**, and fewer still contain anything having to do with spatialized sound. But this will no doubt change as the multimedia market strives to introduce new products for consumer upgrades of their existing systems.

So What Is 3-D Sound Good For?

If we want to improve the quality and ease of interaction within a human interface, communication system, or office or home environment, it is impossible to ignore the sense of hearing. To control the audio sensations of a human user in a positive way requires examining all aspects of human sensitivity to sound. The fact that audio in the real world is heard spatially is the initial impetus for including this aspect within a simulation scenario. For a desirable audio interface, this fact can be combined along with criteria for eliminating the fatiguing aspects of **noise** and **distortion.** In some circumstances, control over these quality issues can make a human interface a safer environment. While this has been recognized for the visual and physical parameters of human usage (VDT screens and wrist supports, for example), the notion of what constitutes a positively human interfaced audio display—i.e., an audio display **ecology**—has been late in coming to the computer.

3-D sound can contribute to the sense of immersivity in a 3-D environment. One might be able to work more effectively within a VR environment if actions were accompanied by appropriate sounds that seemingly emit from their proper locations, in the same way that **texture mapping** is used to improve the quality of shaded images. The jump in realistic representation between wireframe images and texture-mapped images has helped enormously in the popularization of virtual reality and computer graphics.

Sound provides an important channel of feedback that either can be helpfully redundant to a visual cue, or can provide feedback for actions and situations that are out of the field of view of the listener. This is a use of 3-D sound for **situational awareness.** 3-D sound has the advantage over vision in that multiple virtual sound sources can be synthesized to occur anywhere in the 360-degree space around a listener. With an audio display, the focus of attention between virtual sound sources can be switched at will; vision, on the other hand, requires eye or head movement. This is important especially for **speech intelligibility** within multiple communication channel systems, as will be seen in the discussion of applications in Chapter 5.

Perhaps the largest area for future development of **integrated, spatial audio displays** are in high-stress applications where the sound environment as a whole has historically never been considered as a single problem. For instance, the current implementation of sound displays in commercial airline cockpits is essentially scattered among several types of demands for operating the plane safely and efficiently. Input to the auditory channel has been expanded throughout the history of cockpit design on an "as-needed basis," rather than reflecting an integrated design that examines the entire input to a pilot's auditory system. While monitoring the engine sound was (and still is) an important factor in the beginnings of aviation, over the years we have progressed to have an increasingly complex communications and warning system that must be actively monitored by a pilot and given appropriate response. But while visual instrumentation has received considerable attention and guidelines, sound has remained analogous to a neglected child in light of what could be accomplished.

In an airline cockpit, sounds come from many places and at different volume levels, depending on their function: radio frequencies are heard over speakers or monaural headsets; auditory warnings can come from a variety of locations, including speakers mounted near the knees; intracockpit communication comes from the person speaking. These signals must be heard against a background of ambient noise, whose masking effect can affect fatigue and the possibility of operator error. Furthermore, the overall fidelity of spoken auditory warnings is low, compared to the "high fidelity" of the ambient noise. Finally, warning signals tend to reflect mechanical alarms developed nearly half a century ago, rather than using sounds whose aural content better matches the semantic content of the situation. The reasons for this state of auditory chaos are that each audio subsystem is designed separately, the conventional methods for auditory display have been maintained in spite of technological improvement, and there has been a lack of applications research into advanced auditory displays.

In many otherwise technologically sophisticated nations, a *double standard* exists for audio as it reaches the human ear, between the communication system and entertainment industries. The simplest example has to do with **intelligibility.** Simply put, the level of **quality** necessary to make an

	AUDIO PRODUCTION STUDIO	SPACE SHUTTLE LAUNCH CONTROL ROOM
Factors driving standard of audio quality	immersion; ability to attract customers with state-of-the-art technology; realism of virtual audio; reconfigurable	intelligibility; safety; cost; dispersed component design; hardware reliability; replacement; time-proven performance
Frequency range	human hearing (20 Hz–20 kHz)	intelligible speech (200 Hz–5 kHz)
Typical dynamic range of signals	≈40–60 dB (music)	≈10–20 dB (compressed speech)
Noise floor from audio hardware	typically 60 dB, theoretically 90 dB below signal	typically 20, theoretically 40 dB below signal
Environmental context	quiet, acoustically treated rooms, separated by function; limited access	one large, noisy, reverberant room with many people
Relationship of multiple audio sources	integrated within a single audio display design	dispersed between many locations and different devices
Transducers	high-quality headphones or loudspeakers	typically, low-quality communication headsets

TABLE 1.1. Communication systems compared: an audio production studio for music recording and production, and a space station launch facility.

auditory or visual presentation intelligible is far below that used within the entertainment industry. Table 1.1 illustrates the **entertainment–communication system dichotomy** using a professional music recording studio and a NASA space launch facility as examples. Both are high-stress environments, but use radically different approaches to the design of auditory displays.

Part of the reason sound took so long to catch on in human-machine interfaces has to do with the history of low-quality sound implementations in everyday consumer items such as automobiles, kitchen appliances, and computers. Frequently, simple synthesized tones or speech messages are added to an already completed design as a "special feature." Almost every automobile driver has at one time or another wanted to take a knife to the wire connected to the "door open" or "key in ignition" buzzer. An appliance manufacturer has several prototypes of talking ovens and washing machines, but they will probably never be manufactured commercially due to overwhelming rejection in pilot studies. And one frequently hears the anecdote about how most of the executives at one large computer manufacturer had their computer's sound turned completely off, in spite of the fact that their computer featured one of the more interesting interfaces for using sound available at that time. In all of these examples, the added sound is not essential for the operation of the device, and its lack of variation causes annoyance to the consumer after a short period of use. On the design end, the time allotted to implementation is minimized, frequently taken up as a "passing hobby" of a nonexpert on the design team.

In order to be truly successful, the design of an audio interface requires the consultative expertise of a sound specialist, such as an acoustician, musician or psychoacoustician. This expertise is needed to avoid the following common mistakes made in designing speech and audio signals for human-machine interfaces: (1) the sounds are frequently **repetitive, loud,** and **simple,** without any way for the user to change the sound; (2) the sounds are generally used as alarms, instead of as carriers of a variety of information; (3) the quality of the sound is **low**, compared to the CD-entertainment standard, due to the limits of hardware specified within the design; and (4) sound usually contributes little to the ability to use the interface effectively, partially due to the lack of anything but the simplest type of sensor *input*. The automobile horn is an example of a successful acoustic human-machine interface operated by a simple sensor; it responds to the on-off condition of a button on the steering column and is designed to function as a type of alarm. But the *input* to the sensor is anything but simple. Note that most drivers use their horns in different ways, depending on the context: for example, tapping the button briefly can mean "hello," but holding the button down expresses great annoyance. The study of automobile horn "language" as practiced by the world's drivers might be more useful than first expected!

Surround versus 3-D Sound

This book advocates the use of headphones for 3-D sound audition under most circumstances. What about virtual spatial listening over loudspeakers? Simply put, the problem with using loudspeakers for 3-D sound is that control over perceived spatial imagery is greatly sacrificed, since the sound will be reproduced in an unknown environment. In other words, the room and loudspeakers will impose unknown nonlinear transformations that usually cannot be compensated

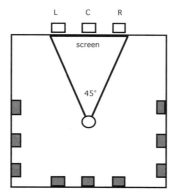

FIGURE 1.9. Dolby speaker arrangement for stereo optical 35mm release. The speaker labeled C is used for dialogue; speakers L and R at left and right front are used to route effects and music, and the shaded speakers are for surround sound effects (after Allen, 1991).

for by the designer or controller of the 3-D audio system. Headphone listening conditions can be roughly approximated from stereo loudspeakers using a technique called **cross-talk cancellation**, described in Chapter 5. But for many applications, loudspeakers are impractical for producing virtual acoustic images.

In the commercial world, there is frequent confusion between devices that attempt to produce 3-D virtual images and those that are in actuality **surround sound** or **spatial enhancement systems.** Intended for music, these systems claim to improve a normal stereo mix, via a process as simple as including an "add-on" box to the home stereo system, or even as a switch on some "boom boxes." Because the art of stereo mixing for loudspeakers involves a fairly carefully created product intended for normal reproduction, these devices in actuality end up destroying spatial and timbral aspects of the original stereo mix. Simply, the intent of a surround sound system or spatial enhancer is to make normal stereo sound more spacious than stereo, albeit by eliminating the control of the recording engineer to predict a final result. If one wires a center loudspeaker out of phase in-between two normally placed loudspeakers, an idea of the intended effect of most of these processors can be experienced.

A further distinction needs to be made between surround sound and **theater surround sound systems.** The setup used by Dolby Laboratories for a 35mm optical stereo sound track is shown in Figure 1.9. By means of Dolby's system, it is possible to derive four channels: center, left, and right channel speakers behind the screen, and multiple "surround" channels to the rear. Most films route dialogue to the center channel, even when people talk from left or right, primarily to avoid the tone coloration problems and synchrony problems between a source's visual and aural location that can occur for an audience of people in different locations. Music and Foley effects (e.g., footsteps outside the width of the picture) use the left and right channels, but rarely the multiple surround channels. The surround channels, which are usually one channel, use ambiance material ("background" sounds that help establish the environmental context) that has no other spatial requirement than to be diffused and enveloping for a large theater audience.

CHARACTERISTICS OF SOUND SOURCES

Describing Waveforms

Before describing how sounds are heard in space, a brief review of basic concepts is presented for those unfamiliar with terminology related to sound and psychoacoustics. It must be emphasized that the following discussion constitutes only "a small part of the story" of the nature of acoustic phenomena and spatial hearing, since the unique features and idiosyncrasies of different listeners and environmental contexts will alter both physical and perceptual features of a spatially audited sound source.

Periodic and Aperiodic Waveforms

Classical textbooks have traditionally distinguished between two categories of sound fields: those that have a **periodic** waveform, and those that have no identifiable periodicity. The former can be described in terms of **frequency** and **intensity**, while the latter, **noise**, can be described statistically. The simplest type of periodicity is inherent within a **sine wave**, which exists either in pure form only in the domain of electronically produced sound, or in theory. When the sine wave is plotted as shown in Figure 1.10, the **intensity** of the waveform on the vertical y axis changes as a function of time on the horizontal x axis. Figure 1.11 shows how the periodic section of the waveform—measured as the number of oscillations over a given time unit of 1 second—determines its **frequency** in Hertz (abbreviated Hz). Intensity can be described in arbitrary units (e.g., ±1) for computational purposes, but once the sound is transduced into the air, the absolute **sound pressure level** in dB SPL is more relevant. Figure 1.11 shows another important concept for sound sources that is particularly relevant to spatial hearing: the relative timing, or **phase** of the waveform. The two sine waves in the figure are identical, except that one is delayed by a period of time equivalent to a quarter period (90 degrees) of the wavelength, which is .001 seconds, or 1 millisecond (abbreviated msec).

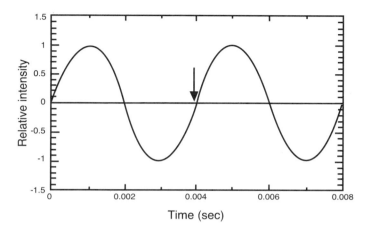

FIGURE 1.10. The simplest waveform: the sine wave. The x axis indicates time in seconds; the y axis, relative intensity in arbitrary units. The number of repeated oscillations per unit time indicates the **frequency**, measured in **Hertz** (abbreviated Hz). The arrow points to where the oscillation repeats itself at .004 seconds; this indicates a frequency of 250 Hz.

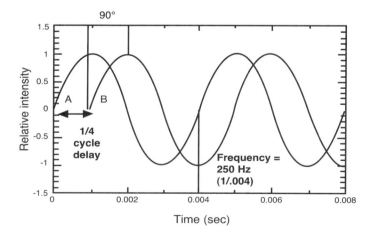

FIGURE 1.11. Two 250-Hz sine waves, A and B. Wave B is delayed by a quarter of a cycle (90°), i.e., by .001 seconds (or 1 millisecond).

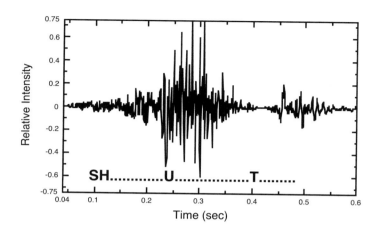

FIGURE 1.12. A waveform plot of the spoken word "shut." A noise-like (aperiodic) portion for the "sh" sound precedes the more pitched "u" sound, while the "t" is transient.

The distinction between periodic and noise vibration becomes artificial when examining many natural sound sources. For instance, speech is comprised of both noise and periodic vibration components. Figure 1.12 shows both noise components (*sh*) and quasi-periodic components (*u* and *t*) within a speech recording of the word "shut." A noisy source can also be amplitude modulated to have a frequency of its own (known as its **modulating frequency**). Figure 1.13 shows at top a noise waveform and a triangle waveform; multiplying the two together results in the amplitude-modulated noise waveform shown below them.

Sound sources can also transmit frequencies below and beyond the range of human hearing (i.e., < 20 Hz or > 20 kHz). These are **subsonic** and **supersonic** frequencies, respectively. Subsonic frequencies can be felt in terms of tactile sensation; vibration of the body can also occur, and in many contexts can be an annoyance or even a health hazard. Subsonic sounds can also be used to emphasize the character of a sound source, as done in cinematic contexts to suggest disasters such as earthquakes. Commercial products have also existed that transpose bass frequencies from a standard audio system into subsonic vibration, allowing one to "feel" the bass. Regarding supersonics, some preliminary research has shown that humans are capable of detecting frequencies as high as 40 kHz! However, this is not "hearing" as understood in terms of activating hair cells in the inner ear; rather, resonances are formed in the parts of the ear that have more to do with the vestibular system's balance organs than with the basilar membrane.

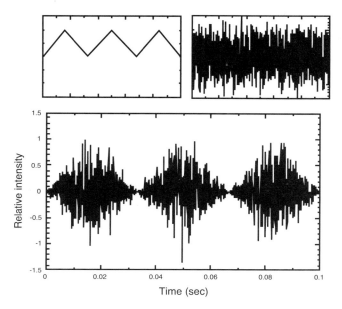

FIGURE 1.13. Top: a triangle wave and a noise waveform. Bottom: the two when multiplied together, i.e., the resulting amplitude modulation.

Digital Sound and Signal Processing

The technological context for implementing 3-D sound is directly tied to both the evolution of **digital signal processing (DSP)** chips for the defense industry, and the commercial introduction of digital audio in the music industry. It was not so long ago that digital representation and manipulation of sound was confined to specialized laboratories. Since their introduction in the United States in 1983, it took only eight years for digital audio technology in the form of the compact disc (CD) to penetrate 28 percent of households (Pohlmann, 1992). Simultaneously, the development of DSP chips, which are capable of millions of computational instructions per second, have over time been increasingly used in everyday audio and graphics applications, and not merely for missile guidance or radar systems.

Analog spatial manipulation devices such as the panpot, joystick, and linear fader are bound to exist for some time. Nevertheless, 3-D sound development is almost certain to remain within the digital domain, due to the sheer volume and complexity of the information that must be analyzed and processed. Changing the character of an input signal in the digital domain to create virtual spatial hearing experiences has been greatly facilitated by the development of DSP chips, particularly since they can be configured as **filters** that modify sound in the same way as in natural spatial hearing. Other advantages of DSP include relatively smaller hardware size, greater predictability of noise, and ease of reconfiguration—one can write different software specifications to the signal processor, rather than resoldering hardware components. DSP is covered more thoroughly in Chapter 4.

In a digital sound system, the highest frequency that can be used will be slightly less than half the **sampling rate** since at least two points are necessary for representing a single oscillation. The "vertical granularity" on the other hand relates to the **quantization** of the signal. The sound pressure level (dB SPL) will depend on the level of amplification used to play back the converted signal; in Figure 1.14, the range ±1 is used on the vertical axis to denote maximum amplitude, which might be represented digitally using 16-bit **signed integer** representation as numbers within a range of –32768 to 32767. Note that the magnitude of a particular sample or dB VU (what you see on the tape recorder) corresponds to no particular intensity at the eardrum; one would need to measure the dB SPL using a sound level meter to have an actual measurement of intensity as it reaches the listener in an actual situation.

Harmonics

While a sine wave is considered technically to be a "simple" waveform, almost all waveforms in nature are in actuality "complex." This means that whatever portion of the waveform we're examining will contain **spectra** in the form of **harmonic** and/or **inharmonic** frequencies, i.e., vibrations that are related to

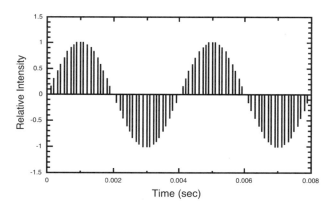

FIGURE 1.14. Sampled version of the sine wave shown in Figure 1.10. At a sampling rate of 10,000 Hz, 40 samples per cycle of a 250-Hz sine wave would be obtained.

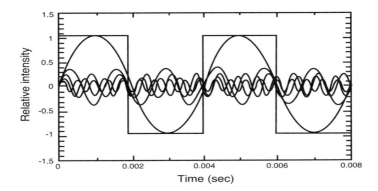

FIGURE 1.15. A square wave superimposed over sine waves that represent its first five odd-numbered ratio harmonics (250, 750, 1250, 1750, and 2250 Hz). An actual square wave would be composed of successive odd-numbered harmonics up to the highest frequency capable of being produced by the audio system (in a digital system, almost half the sampling rate).

the **fundamental frequency** of a vibrating medium by ratios that are either mathematically simple (e.g., 3:2) or complex (e.g., 3.14159:2). Most complex waveforms in nature contain many harmonic and inharmonic frequencies, while waveforms containing *only* harmonically related tones are almost always synthetic. For example, on a synthesizer, one can produce a **square wave** (see Figure 1.15), which is composed of a fundamental frequency with harmonics at odd numbered multiples. The intensity of each harmonic is the reciprocal of the harmonic relative to the fundamental: e.g., given a fundamental at 250 Hz, the first harmonic at 750 Hz will be one third the intensity, the second harmonic at 1250 Hz will be one fifth the intensity, etc. Figure 1.16 shows another way of representing the harmonics of the square wave; the x axis is used for indicating the frequency of each harmonic, instead of time.

Fourier Analysis

We can look at the harmonic content of a sound source over a single period in time by performing a **Fourier analysis** on the waveform. Fourier analysis is implemented for a finite-length digital sequence in terms of the **discrete Fourier transform** (**DFT**), and the DFT is often implemented in terms of the more computationally efficient **fast Fourier transform** (**FFT**). The phrase "Fourier transform" is used here to refer to either technique. The ultimate goal is to obtain a **spectral analysis** of a sound source, in terms of a two-

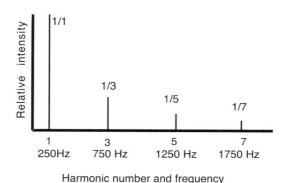

FIGURE 1.16. A plot of the relative intensity for each harmonic component of a square wave up to the seventh harmonic.

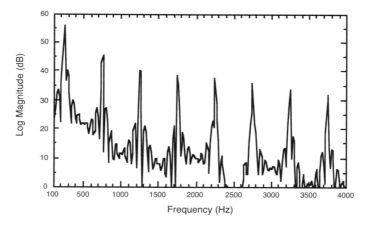

FIGURE 1.17. Fourier transform of the square wave shown in Figures 1.15 and 1.16.

dimensional frequency-versus-dB magnitude plot. This is useful for determining, for example, the effects of filtering that results from the outer ear, or from a surface within an environmental enclosure that a sound may reflect against.

The Fourier transform works for sound analysis by mathematically decomposing a complex waveform into a series of sine waves whose amplitudes and phases can then be determined. Figure 1.17 shows the FFT of

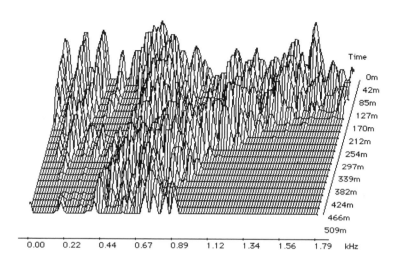

FIGURE 1.18. A series of FFTs taken over the temporal evolution of a waveform shows the evolution and diminution of various harmonic components from 0–1790 Hz.

the square wave shown in the previous figures; note that time is now absent from the picture. It should be noted that Fourier analysis works perfectly only when an "infinite" waveform is analyzed. Usually, we need to **window** a portion of the waveform, which produces errors seen in the "side lobes" of the peaks of Figure 1.17. If it worked perfectly, a picture similar to that shown in Figure 1.16 would result instead.

The fact that a DFT or FFT essentially "freezes time" makes it less useful for analyzing the temporal evolution of a sound source, although it is usually perfectly adequate for describing a filter or loudspeaker frequency response. In order to get an idea of the temporal evolution of the harmonic structure of a sound, multiple FFTs of a waveform can be taken over successive portions of time, and then arranged on the z axis of a 3-D graph, as shown in Figure 1.18.

Each slice in time reveals particular spectral **resonances**, also called **spectral peaks**. These are locations in the overall spectrum of a sound where

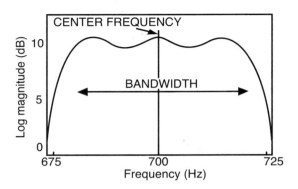

FIGURE 1.19. Close-up of a resonance in Figure 1.18. An **anti-resonance** would appear as an "upside-down" version of a resonance.

energy is greatest. Each resonance can be described in terms of its **center frequency**, which can be thought of as the "middle location" of the resonance, and a **bandwidth**, the extent of the resonance across frequency.

For instance, in the foreground of Figure 1.18, there is a resonance with a center frequency of around 700 Hz, with a bandwidth of about 50 Hz; a zoom picture of this single resonance is shown in Figure 1.19. An **anti-resonance**, or **spectral notch**, can be thought of as the "opposite" of a spectral peak; it appears as a "dip" in the spectrum and can also be described using bandwidth and center frequency terminology. Bandwidths and center frequencies are frequently used to describe the resonances and anti-resonances of either natural systems, such as the outer ear, or of filters, such as those used to simulate the outer ear.

Fourier transforms are one of the most useful mathematical tools for analyzing sound, particularly for designing and analyzing the digital filters that constitute the bulk of the signal processing used in 3-D sound. Many relatively inexpensive software packages allow Fourier analysis of digitized waveforms, without the need on the part of the user to understand the mathematics behind the transform. But one should realize that other transforms and means for signal analysis do exist, each with particular advantages and drawbacks. While it is worthwhile to attempt a mathematical understanding of the FFT and DFT (see Chapter 6 for some helpful resources in this area), it is perhaps even more important to realize that a Fourier transform provides an abstract, repeatable representation of a waveform's harmonic structure—but that this representation will differ as a function of the detail of the analysis (a function of computational resources), the type of windowing used, and the temporal duration of the waveform analyzed.

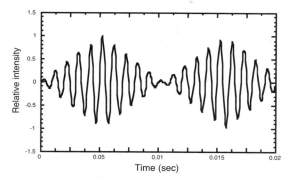

FIGURE 1.20. The amplitude modulation of a 100 Hz sine wave by a 10 Hz triangle wave on a synthesizer results in a constant, synthetic amplitude envelope.

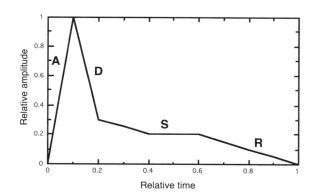

FIGURE 1.21. The **ADSR** (**A**ttack-**D**ecay-**S**ustain-**R**elease) amplitude envelope, a simplified way of describing the overall intensity of a complex sound over time. It is commonly used on sound synthesizers and samplers to describe what occurs with a single keystroke, or digital "note on" command. Specifically, the ADS portion of the sound is what happens when a key is pushed down (note on), and the R is what occurs when the key is let up (note off).

Amplitude Envelope

The concept of a waveform's **amplitude envelope** refers to the overall shape of a waveform's intensity over time, usually with reference to natural acoustic phenomena. Figure 1.20 shows the amplitude modulation of a sine wave by two cycles of a triangle wave, such as might be produced using two electronic waveform **oscillators** on a synthesizer. In this case, the triangular-shaped amplitude modulation is equivalent to the amplitude envelope, and will remain constant as long as the oscillators are not switched off. But most naturally-occurring acoustical phenomena involve the repeated activation and decay of a vibratory medium, particularly musical instruments. This causes the overall amplitude to build to a maximum relatively quickly and then decay relatively slowly, in a manner characteristic of the particular sound source.

Figure 1.21 shows an "overall" amplitude envelope of all the harmonics of a complex waveform, as commonly specified on a tone generator interface to synthesize a natural sound. Note that, like a piano, the sound does not stop when the key is released, but takes a brief moment to "die out." In reality, each individual harmonic and inharmonic component of a natural, complex sound will have its own amplitude envelope, making the *overall* amplitude envelope only a rough approximation of the sound's temporal evolution. To experience this yourself, play the lowest note of a grand piano once while holding down the sustain pedal. Over the period of a minute or so, you should have heard several very different amplitude envelopes at work, emphasizing and de-emphasizing different harmonics over time. It is the complex interaction of these harmonics over time that gives the grand piano its special quality and makes it very difficult to synthesize. Sound designers must work very hard to avoid regularity in the short term or long term envelopes of each of a synthesized sound's harmonics as well as its overall amplitude envelope, if a "natural" as opposed to "synthetic" character is desired.

PERCEPTION

Psychoacoustics and Applications

The evaluation of 3-D sound from the perspective of reporting the experience to other people involves another level of transformation in the source-medium-receiver model. Changes in the source or medium can potentially result in changes in the spatial imagery heard by the receiver. However, for the receiver to describe the change in spatial imagery to others, a descriptive language must be used to describe their percepts. While a physiological feature like blood pressure can be measured directly, psychophysical investigations must deal with mental processing, and can only indicate probabilities of human response to a stimulus.

Similarly, research into the physical aspects of sound allows determination of variables that are usually very predictable. For instance, given knowledge of the properties of a particular volume of air, one can predict the speed of sound with a great deal of consistency. On the other hand, because humans tend to be less consistent (between each other or within themselves), a distinction is made between the physics and **psychophysics** of sound (termed **psychoacoustics**). Statistical techniques and carefully-designed, repeatable experiments are therefore absolutely necessary for describing the nature of auditory spatial perception.

One problem that is particularly difficult for spatial hearing research involves determining a language (or **response variable**) that has as great a consistency as possible among subjects. The choice of appropriate response variable is highly important, as will be seen in the comparison of psychoacoustic studies of spatial hearing in Chapters 2 and 3. Marketing and advertising people have more devious ways of manipulating descriptions, but statements like "I feel *good* about my car" can mean something completely different to many people, as can "experiencing new and exciting dimensions of cyber-audio space."

Psychoacoustic experiments work under acoustic conditions that are not at all similar to conditions found in natural spatial hearing, in order to limit the number of uncontrolled variables. The problem for 3-D sound is that, usually, the more control an experimenter obtains, the less the situation reflects reality. For instance, the majority of spatial hearing studies have used an anechoic chamber as the environmental context, in spite of the fact that reverberation can have a significant effect on spatial perception. The types of sounds utilized pose additional limitations, which typically are one or more of the following: pure tones (sinusoids), noise (either full spectrum or filtered), and clicks (impulses). The use of noise more closely approximates the complex sounds of the real world, while clicks simulate the onset transient information of a complex sound. Speech sounds have also been used in many experiments. Hence, the constraints of experimental designs must be taken into account for any application that extends from them. The conclusions of psychoacoustic research related to spatial hearing are relevant to the extent that they elucidate specific mechanisms of localization and indicate abilities of these mechanisms under controlled conditions.

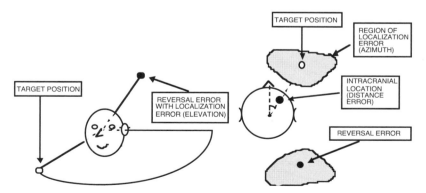

FIGURE 1.22. Types of localization error reported in the literature and discussed later in this book. Left side: frontal view. Right side: perspective view. Reproduced with permission from Human Factors, Vol. 35, No. 2, 1993. Copyright 1993 by the Human Factors and Ergonomics Society. All rights reserved.

Another important feature necessary for appreciating a particular psychoacoustic experiment's applicability is to distinguish between ones where subjects were asked to estimate the location of a virtual image (Blauert, 1983). This problem can be seen in cases where a subject is asked to identify which of a set of sources is producing a sound (e.g., Fisher and Freedman, 1968). In these cases, the available choices are limited to what the experimenter has chosen. Localization estimates will therefore differ from the actual perception of where the sound is located for the listener.

The characterization of localization estimates—where a listener hears a virtual acoustic source, as opposed to its intended position—is shown in Figure 1.22. The "target" position refers to the position of the sound source where the spatial cues were measured. The perceptual deviation from the intended target is measured in terms of localization error (the absolute error in estimating azimuth and elevation), reversals (azimuth errors between the front and rear hemispheres), and distance errors (in particular, hearing the sound inside the head).

Perceptual Correlates of Frequency, Intensity, and Spectral Content

The basic physical descriptors of a sound event—frequency, intensity, spectral content, and duration—have corresponding perceptual terminology. For instance, the corresponding perceptual term for frequency is **pitch.** For a sine wave, the inner ear converts temporal vibrations of the eardrum to a particular point along the basilar membrane (the **place theory** of pitch), in such a way that a doubling of frequency (an **octave** relationship) corresponds to a doubling of the distance.

The equivalence between pitch and frequency can become messy with real sound sources. For example, musicians commonly modulate frequency over time using a technique known as **vibrato.** Although the frequency is varied as much as a semitone at a rate of 5–8 Hz, a single pitch is perceived. Consider the technique of **stretch tuning** used by piano tuners; specifically, the relationship between the fundamental frequency of a piano string and pitch as associated with the tempered piano scale. As one goes up from the reference note A 440, a professional tuner will progressively raise the frequency of the strings and will tune strings progressively flat for lower pitches. Finally, the phenomenon of **fundamental tracking** explains how one can hear the pitch of a bass player on a 2-inch radio speaker incapable of reproducing that frequency. In most cases, the ear determines the pitch at the correct fundamental based on the harmonic relationship of the higher harmonics that the speaker is capable of reproducing.

Loudness is the perceptual correlate of intensity. The loudness of a waveform varies widely as a function of frequency. For sine waves, equivalent loudness contours can be looked up on an **equal loudness** (also known as a **Fletcher-Munson**) graph. The curves on this graph establish that, for instance, a tone at 60 dB SPL is not as loud at 200 Hz as it is at 4000 Hz. Not surprisingly, the frequency content of speech (approximately 100 Hz–7 kHz) is within the part of the curves that show maximal sensitivity. The loudness and "bass boost" buttons found on consumer audio equipment are designed to compensate for this decrease in sensitivity to low frequencies at low sound pressure levels.

The spectral content (spectrum) of a sound source, along with the manner that the content changes over time, is largely responsible for the perceptual quality of **timbre.** Sometimes referred to simply as "tone color," timbre is actually difficult to define. A "negative" definition that is commonly cited for timbre is "the quality of sound that distinguishes it from other sounds of the same pitch and loudness." This could be extended to spatial hearing, in that two sounds with the same pitch, loudness, and timbre will sound different by virtue of placement at different spatial locations. Regarding timbre, the following five parameters, taken from the work of Schouten (1968), seem best to distinguish sounds that are otherwise the same: (1) the range between tonal and noise-like character; (2) the spectral envelope (the evolution of harmonics over time); (3) the time envelope in terms of rise, duration, and decay; (4) small pitch changes in both the fundamental frequency and the spectral envelope; and (5) the information contained in the onset of the sound, as compared to the rest of the sound.

The perception of timbre was explained early on by Helmholtz (1877) as depending on the presence and intensity of harmonics within a complex sound. Timbre could then be explained as the time evolution of energy within the harmonics contained within a particular sound "object," as shown by the Fourier transform in Figure 1.18. But in later research, the integration of sound energy at various harmonics was found to be better explained by the role of **critical bands** (Zwicker, 1961; Zwicker and Scharf, 1965), which are important for explaining many psychoacoustic phenomena, including loudness and the perception of phase.

The frequency range of a critical band is related to the relative sensitivity of activated regions on the basilar membrane (see, e.g., Roederer, 1975, or Scharf, 1970). A full explanation of the basic theory is not attempted here, but it is important to realize that the spectral features shown by Fourier transforms have psychoacoustic equivalents, just as frequency and intensity are related to pitch and loudness. Critical band theory is usually simplified to mean the following: a complex sound is analyzed by the ear with a bank of 24 filters, each tuned with a successive center frequency and bandwidth so as to cover the entire audio range (just like a graphic equalizer). The size of the critical band approximates a 1/3 octave bandwidth; harmonics falling within a critical bandwidth will be integrated in such a way that the strongest harmonic within a particular band will mask ("drown out") other harmonics within that same band, more so than if these other harmonics were in other bands. This suggests that the auditory system is capable of analyzing an incoming waveform within each critical band independently. A summation of energy within each critical band is useful for determining, for instance, overall loudness, or speech intelligibility, and there is evidence that some aspects of spatial hearing involve a comparison of energy within critical bands at the left and right ears. Table 1.2 gives the center frequency and bandwidth for each critical band.

This explanation of critical bands, while widely used, in fact oversimplifies the scientific evidence for their existence and behavior. In spite of its widespread

Critical band number	Center frequency Hz	Bandwidth Hz
1	50	80
2	150	100
3	250	100
4	350	100
5	450	110
6	570	120
7	700	140
8	840	150
9	1,000	160
10	1,170	190
11	1,370	210
12	1,600	240
13	1,850	280
14	2,150	320
15	2,500	380
16	2,900	450
17	3,400	550
18	4,000	700
19	4,800	900
20	5,800	1,100
21	7,000	1,300
22	8,500	1,800
23	10,500	2,500
24	13,500	3,500

TABLE 1.2. Critical band center frequencies and bandwidths (after Zwicker, 1961).

citation in the development of audio compression schemes such as the MPEG (Motion Picture Expert Group) standard (ISO 1993), any meaningful application of the theory is made difficult by the fact that the size of a critical bandwidth (as originally defined) is dependent both on the pressure level and the spatial orientation of the sound to the listener. While we might be able to determine the latter through the use of a virtual acoustic system, the sound pressure level is impossible to predict at a listener's ear within a given situation, if for no other reason than the fact that the user can change the volume control level.

Cognition

Cognition refers to a point of view in perceptual psychology that stresses not only sensation, but also "higher-level" processes of memory, understanding and reasoning. Like the phenomenon of visual capture discussed earlier, familiarity and expectation can modify spatial judgments, circumventing any prediction of spatial location that may have been made strictly on the basis of localization cues. A famous 3-D sound system demonstration tape featured two examples that still amaze those new to virtual acoustics. The first example is the sound of scissors cutting hair, as if very near your ear. The simulation is convincing enough to have had some listeners feeling their head to make sure that someone wasn't actually giving them a trim. The second example is of lighting a cigarette and drinking a glass of water. The experience of the sound of the water and cigarette as being near your mouth is very vivid; in particular, the sound seems to actually be in front of you, a difficult illusion to achieve (see the discussion in Chapter 2 on reversals). The producer of the demonstration cites these examples as proof of the quality of their sound spatialization system.

Truthfully, given our knowledge of the circumstances in which these particular sounds have been experienced, it would be difficult to imagine those sounds coming from anywhere else. This is even true when the sounds are heard *without* 3-D sound processing! For instance, you have probably not lit a cigarette or drank a glass of water from the back of your head, so it's natural to hear the sounds as coming from the front. Similarly, you can impress an uninitiated listener if you make a monaural recording with an ordinary microphone of hair being cut, and then play it back over headphones at the same SPL as it would be heard in real life.

The moral to this is that the choice of sound sources and consideration of the listener's associations and experiences are important factors in the design of virtual acoustic displays. Unfortunately, the determination of a metric for something as multidimensional and subjective as "familiarity" for an arbitrary sound source is so difficult that it defies implementation into the engineering requirements of a 3-D sound system. These are factors that determine the level of art involved in the *application* of virtual acoustics by the end user.

CHAPTER TWO

Overview of Spatial Hearing

Part I: Azimuth and Elevation Perception

INTERAURAL TIME AND INTENSITY CUES

Lateralization

The most important cues for localizing a sound source's angular position involve the relative difference of the wavefront at the two ears on the horizontal plane. From an evolutionary standpoint, the horizontal placement of the ears maximizes differences for sound events occurring around the listener, rather than from below or above, enabling audition of terrain-based sources outside the visual field of view. The frequency-dependent cues of **interaural time differences** (abbreviated **ITD**) and **interaural intensity differences** (abbreviated **IID**) are reviewed in terms of physical and perceptual attributes in this chapter. These differences have been studied extensively in the psychoacoustic literature and are sometimes referred to as the "duplex theory" of localization (Rayleigh, 1907).

In order to describe difference cues, psychoacoustic experiments are constrained according to a **lateralization** paradigm. These experiments involve the manipulation of ITD and IID in order to determine the relative sensitivity of physiological mechanisms to these cues. Although lateralization can occur with speakers in anechoic environments, lateralization experiments almost always use headphones. The word "lateralized" has therefore come to indicate a special case of localization, where:

1. the spatial percept is heard inside the head, mostly along the interaural axis between the ears; and
2. the means of producing the percept involves manipulation of interaural time or intensity differences over headphones.

Within a lateralization paradigm, it is possible to make limited but provable hypotheses about the physiology of the auditory localization system by manipulating simple but controllable parameters; for instance, frequency and relative interaural time of arrival (interaural phase) of sine waves or band-pass noise as independent variables. As a result, a number of studies have reached established conclusions regarding auditory localization from different restricted perspectives.

The experimental designs have focused generally in three areas. First are measurements of what are called "just noticeable differences" (**jnds**) for different interaural conditions and stimuli; related to this are studies of **minimum audible angles** (**MAA**). Second are studies of the interaction between the cues, especially the manner in which one cue cancels the effect of the other (**time-intensity trading**). The third class of studies involves assessments of the overall efficacy of interaural cues, including perceptual descriptions of the mapping of displacement within the head. The latter area is of primary interest for applications of what can be termed "two-dimensional localization" within a headphone-based auditory display (Middlebrooks and Green, 1991).

Physical Basis of Lateralization

Lateralization mimics the interaural differences present in natural spatial hearing. To see this, consider the hypothetical situation of a seated listener with a perfectly round head and no outer ears, placed at a distance in an anechoic chamber from a broadband sound source at eye level (see Figure 2.1). Modeling this situation involves calculating two paths representing the sound source wavefront from its center of origin to two points representing the entrance to the ear canal. An additional simplification of this model is the placement of these points exactly at the midline crossing the sphere, at the ends of the interaural axis. With the source at position **A** at 0 degrees azimuth, the path lengths are equal, causing the wavefront to arrive at the eardrums at the same time and with equal intensity. At position **B**, the sound source is at 60 degrees azimuth to the right of the listener, and the paths are now unequal; this will cause the sound source wavefront to arrive later in time at the left ear relative to the right.

The sound path length difference just described is the basis of the interaural time difference (ITD) cue, and it relates to the hearing system's ability to detect interaural phase differences below approximately 1 kHz. A value around .00065 sec is a good approximation of the maximum value for an average head. The

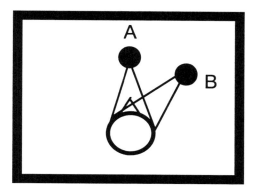

FIGURE 2.1. A listener in an anechoic chamber, with a sound source oriented directly ahead on the median plane (A) and displaced to 60 degrees azimuth.

sound source at position **B** in Figure 2.1 will also yield a significant interaural intensity difference (IID) cue, but only for those waveform components that are smaller than the diameter of the head, i.e., for frequencies greater than about 1.5 kHz. Higher frequencies will be attenuated at the left ear because the head acts as an obstacle, creating a "shadow" effect at the opposite side. The relation of a wavefront to an obstacle of fixed size is such that the shadow effect increases with increasing frequency (i.e., decreasing size of the wavelength).

For instance, a 3-kHz sine wave at 90 degrees azimuth will be attenuated by about 10 dB, a 6-kHz sine wave will be attenuated by about 20 dB, and a 10-kHz wave by about 35 dB (Feddersen, et al., 1957; Middlebrooks and Green, 1991). But below approximately 1 kHz, IID is no longer effective as a natural spatial hearing cue because longer wavelengths will diffract ("bend") around the obstructing surface of the head, thereby minimizing the intensity differences. Everyday examples of diffraction are how sunlight bends around the horizon even after the sun has set; or how, between two office cubicles, low frequencies caused by a source such as a printer are less effectively masked than the relatively higher speech frequencies of a telephone conversation. This is also why sufficient analysis of community noise complaints should include potential paths of low-frequency energy (from sources such as building machinery or automobile sound systems) for developing sound-proofing treatment solutions.

Independent of frequency content, variations in the overall difference between left and right intensity levels at the eardrum are interpreted as changes in the sound source position from the perspective of the listener. Consider the primary spatial auditory cueing device built into stereo recording consoles, the **panpot** (short for "panoramic potentiometer"). Over headphones, the panpot creates interaural level differences without regard to frequency, yet it works for separating sound sources in most applications. This is because the frequency content of typical sounds includes frequencies above and below the hypothetical "cut-off" points for IID and ITD, and that listeners are sensitive to IID cues for lateralization across most of the audible frequency range, down to at least 200 Hz (Blauert, 1983).

ITD Envelope Cue

Earlier research into spatial hearing found results that suggested that the duplex theory—IID and ITD cues—operated over exclusive frequency domains (e.g., Stevens and Newman, 1936). Stated simply, the theory was that one neurological component for spatial hearing responded to ITD cues only for "low frequencies" below 1.5 kHz, and the other component evaluated IID cues for the "high frequencies" above this cut-off point. But later research has revealed that timing information can be used for higher frequencies because the timing differences in amplitude envelopes are detected (Leakey, Sayers, and Cherry, 1958; Henning, 1974; Trahiotis and Bernstein, 1986). This **ITD envelope cue** is based on the hearing system's extraction of the timing differences of the onsets of amplitude envelopes, rather than of the timing of the waveform within the envelope.

The ITD envelope cue is illustrated in Figure 2.2. **A** and **B** show sine waves at the left and right ears that are below 800 Hz. Because the half period of the waves is larger than the size of the head, it is possible for the auditory system to detect the phase of these waveforms unambiguously, and the ITD cue can function. Above a critical point of about 1.6 kHz, sine waves become smaller than the size of the head, creating an ambiguous situation: the phase information in relationship to relative time of arrival at the ears can no longer convey which is the leading wavefront; i.e., whether **D** leads **E**, or **E** leads **F** in Figure 2.2. But if the sine waves are increased and decreased in amplitude (via a process known as **amplitude modulation**) then an amplitude envelope is imposed on the sine wave (see **X** and **Y** in Figure 2.2).

The envelope ITD cue results from the hearing system's extraction of the timing of the relative onsets of the amplitude envelopes. Note that the frequency of the amplitude envelope (i.e., the **amplitude modulating frequency**) is much lower than the frequency of the sound contained within the envelope (referred to technically as the **carrier frequency**). The auditory system somehow extracts the overall amplitude envelope of higher-frequency components at both ears and measures the difference in time of arrival of the two envelopes. This was proven with experiments using high-pass–filtered noise at various "cut-off" points above 1.5 kHz, supposedly the highest usable frequency range for ITD. These high-passed stimuli still caused displaced spatial images for subjects when interaural timing differences were manipulated.

In normal spatial hearing, the interaural comparison of envelope timing is probably evaluated separately within each critical band and then combined toward

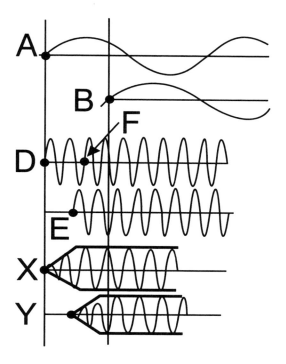

FIGURE 2.2. Transient envelope cue (refer to text for explanation).

forming a final localization judgment. But for the broadband sources typically encountered in normal spatial hearing, the overall efficacy of the cue is probably limited compared to low-frequency ITD and IID (Middlebrooks and Green, 1991); the cue by itself can only move a spatial image along a relatively smaller distance along the interaural axis (Trahiotis and Bernstein, 1986).

Perception of Lateralization

Lateralization illustrates a fundamental example of virtual, as opposed to actual, sound source position. When identical (monaural, also termed **diotic**) sounds are delivered from stereo headphones, the spatialized image appears at a virtual position in the center of the head, rather than as two sounds in either transducer. A similar situation occurs with two-channel loudspeaker systems; a good way to find the "sweet spot" for your home stereo system is to adjust your seating position (or the balance knob) until a monaural radio broadcast sounds like a virtual sound source located at the midpoint between the loudspeakers.

As interaural intensity or time differences are increased past a particular threshold, the position of the virtual sound image will begin to shift toward the

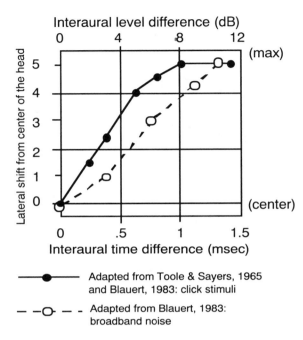

FIGURE 2.3. Perception of interaural differences within a lateralization paradigm.

the ear leading in time or greater in amplitude. Once a critical value of ITD or IID is reached, the sound stops moving along the interaural axis and remains at the leading or more intense ear. Figure 2.3 shows these differences rated by subjects on a 1 to 5 scale, where 5 represents maximum displacement to one side of the head on the interaural axis (adapted from the summary in Blauert, 1983). Depending on the type of stimulus, the effective range of interaural timing differences is approximately between .005 to 1.5 msec, and the effective range of interaural intensity differences is between approximately 1 to 10 dB (Sayers, 1964; Toole and Sayers, 1965). The upper range for intensity differences is more difficult to determine than that for timing. This is because within the upper range of intensity differences, a change in position is difficult to discriminate from a change in auditory extent (Durlach and Colburn, 1978; Blauert, 1983).

Although the literature often makes the claim that lateralized sound sources are heard only along the interaural axis, there is also a vertical component experienced by some listeners—as interaural differences are minimized, the source can travel *upward* (von Békésy, 1960). In addition, lateralized sound is

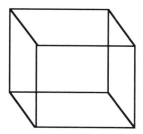

FIGURE 2.4. The Necker cube, a reversible perspective illusion comparable to front–back ambiguity of sound sources.

heard sometimes behind, and sometimes in front of the interaural axis by listeners. Figure 2.4 illustrates the visual equivalent of a reversible percept; the image can be seen (heard) in two different ways—with the cube going either upward on the frontal plane, or downward (there are many other examples of ambiguous percepts, e.g., "Rubin's figure," where facial silhouettes form the "Peter–Paul goblet"). Georg von Békésy related a humorous anecdote regarding these types of auditory reversals. During the early days of radio, one highly placed government official listened to orchestral broadcasts wearing headphones, but complained that he heard the orchestra behind him, instead of in front. His wife was convinced he should see a psychiatrist, until von Békésy showed him that the percept was ambiguous as to its location; it could be interchanged with a location in the center of the head, or in front (von Békésy, 1967).

The Precedence Effect

A lateralization paradigm involving ITDs larger than those predicted by the models discussed previously illustrates the **precedence effect**, first defined by Wallach, Newman, and Rosenzweig (1949). It is also called the **Haas** effect, based on that author's examination of the influence of time delays on speech intelligibility (Haas, 1972); Blauert (1983) terms it "the law of the first wavefront" (see Gardner, 1968, for a historical review). The precedence effect explains an important inhibitory mechanism of the auditory system that allows one to localize sounds in the presence of reverberation.

If one takes stereo headphones or loudspeakers and processes the left channel using a variable delay device, a series of perceptual events will occur akin to those shown in Figure 2.5. The values shown here are only approximate; the level, type of source used, and individual threshold of each listener will alter the actual values. In natural spatial hearing, the angle of incidence of the delayed

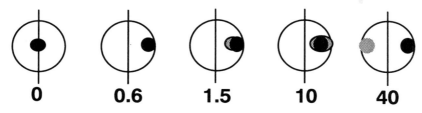

Approximate delay time to left channel (msec)

FIGURE 2.5.   Perceptual effect of increasing ITD from 0–40 msecs. Overhead perspective, with the listener facing downwards (refer to text for explanation).

sound relative to the undelayed sound will give different results as well. Nevertheless, the pattern of transitions characterized in Figure 2.5 applies to many interchannel time-delay effects heard over headphones.

Between 0–0.6 msec, the image shifts along the lateral axis, as predicted by the role of ITD in natural spatial hearing (see also Figure 2.3). In the range of approximately .7–35 msec, the lateralization of the source is still to a position associated with the undelayed channel; it takes precedence over the later channel as far as a determination of the location of the image source. Note that the percept of the 1.5 msec delay shown in Figure 2.5 is in the same location as that for the delay at .6 msec. This is roughly the same position a single sound source would effect at a position corresponding to maximum ITD.

Figure 2.5 also shows a slight change in the image width (shown as the shaded area around the sound source location) at 1.5 msec, at a delay value larger than the maximal ITD value for lateralization; generally, the width will continue to increase over a certain range. Other percepts that change as the time delay is increased can include tone color, loudness, and a slight shift in position toward the delayed channel (Blauert, 1983). At higher delays (marked 10 msec in Figure 2.5) the "center of gravity" of the source can also move back toward the center. At a somewhat ambiguous point that, again, depends on the sound source (marked 40 msec in Figure 2.5), the broadened source will split like an amoeba into two separate, unbroadened images; the latter image is termed an **echo,** familiar from aural experiences in large empty rooms or in natural features such as canyons.

ITD, IID, and Barn Owls: A Neurological Processing Model

It is possible at this point to give a thumbnail sketch of neurological aspects of ITD and IID. Localization can be said physiologically to involve two stages; first, a comparison of the signal arriving at both ears, and second, monaural processing of the signal into auditory space perception. At higher cognitive levels, the spatial percept can be a function of memory, interpretation, association, and both monaural and binaural cues. For the moment, the neurological combining of ITD and IID is overviewed.

Recent insight into auditory localization in humans has been gained by working with barn owls. The work of Knudsen, Konishi, and their colleagues has focused on this owl's use of two-ear listening in the dark as a means of finding its prey (e.g., Knudsen and Konishi, 1978; Knudsen and Brainard, 1991; Konishi, 1993). Their work is valuable in that they have been able to isolate neurologically most of the neurological steps in the formation of a single spatial image from ITD and IID information at the two ears. While it is very likely that there are similarities in the neurological processing of humans and owls, one interesting difference has to do with elevation perception. It turns out that barn owls primarily use ITD for horizontal plane localization, and IID for vertical localization. This is because, unlike humans, the ears of this particular owl are placed asymmetrically; the left ear points downward and the right ear points upward, meaning an additional interaural cue for elevation can be gained in a manner not available to humans. This makes sense from an evolutionary standpoint, considering that owls usually hunt their prey from above, while humans run toward or away from animals at ground level.

Konishi's work also gives insight into how ITD and IID are processed sequentially in the brain. In humans, interaural time differences are thought to be cross-correlated in the medial superior olive of the neurological pathway; some theories propose a single mechanism sensitive to both time and amplitude, while others propose two different physiological mechanisms that are eventually cross-correlated (Roederer, 1975). The model for barn owls analyzed by Konishi (1993) is explained by the simple diagram shown in Figure 2.6; in viewing other neurological research, Konishi suggests that brains follow certain general rules for information processing that are common between different sensory systems and even species.

At the inner ear (**A** in Figure 2.6), hair cells along the basilar membrane create electrical discharges, or "firings," of auditory neurons, that are separated in terms of time of intensity at a fairly low level. Hair-cell firing explains the detection of interaural timing differences well, but the detection of IID based on firing rate is less clear. Most likely, a sort of conversion process happens, based on a frequency-selective process of firing *rates* within critical bands.

In Konishi's model, the intensity is transmitted in terms of the firing rate of selected frequency-tuned neurons, to the *angular nucleus* (**B** in Figure 2.6), while the timing information is extracted by the *magnocellular nucleus* (**C**) into a position within neurons that approximates a digital delay line. ITD is then determined in terms of **coincidence detection** at the *laminar nucleus* (**D**) which occurs by combining information from delay lines at either ear. Coincidence detection, a model first proposed by Jeffress (1948), is explained

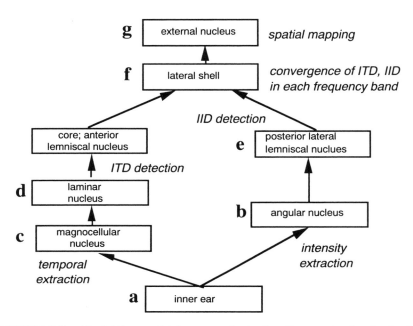

FIGURE 2.6. Konishi's model for neurological pathways in the brain (see text). *Adapted from Konishi (1993) "Listening With Two Ears,"* Scientific American, *vol. 268, No. 4.*

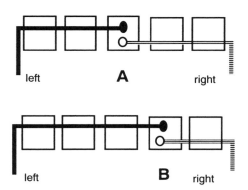

FIGURE 2.7. Simplified illustration of coincidence detection for ITD in the laminar nucleus. Each square represents a unit "time delay"; coincidences of firings at a particular neuron (unit) are mapped to lateral position. In case A (top) a sound from directly in front causes both left and right ear firings to arrive at the central unit; in case B, a sound displaced to, e.g., right 45 degrees azimuth has a coincident firing at a rightward unit.

briefly in Figure 2.7; the elaborations of the model are beyond the scope of the current discussion (see, e.g., Blauert, 1983). A similar combining of IID from both ears by comparing the firing rates occurs at the *posterior lateral lemniscal nucleus* (**E** in Figure 2.6). Remember that all of this occurs simultaneously, yet separately, for different regions for frequency; the information from different frequency channels as well as of the ITD and IID differences occurs at a higher level called the *lateral shell* (**F**), in the midbrain auditory areas.

Perhaps the most interesting result of Konishi's and Knudsen's work was the identification of even higher-level processes at what are called *space-specific neurons* (**G** in Figure 2.6) (Knudsen and Brainard, 1991; Konishi, 1993). It turns out that the owl has specific neurons that only fire for certain directions of sound! Furthermore, neural processing of auditory localization information can be altered as a function of an adaptation to altered *visual* stimuli. Knudsen and Brainard (1991) raised owls with prisms mounted in front of the eyes in order to displace the location of visual sources optically. Their experiments with these owls showed that auditory cues were reinterpreted in order to fit the displaced visual mapping. Hence, the phenomenon of *visual dominance* is suggested, especially since the higher-level mapping of both auditory and visual space occurs at the superior colliculus. This is itself an exciting link between neurological functions and higher-level cognitive activities: in other words, how we determine that some directions are more important than others. Although unproved, a possible extension of the integration of visual and auditory cues into higher neurological maps is to the cognitive mapping of visual and sound events for human orientation and memory. Perhaps similar processes underlie visual cognitive "mapping" (e.g., how a human organizes an unfamiliar city landscape within mental space) and auditory spatial mapping of multiple sound sources (e.g., the sounds heard within the same space) that could be traced to interrelated neurological mapping functions.

HEAD MOVEMENT AND SOURCE MOVEMENT CUES

Head Movement

In everyday perception we use head motion with our aural and visual senses in an attempt to acquire sound sources visually. Therefore, an important source of information for localization is obtained from head motion cues. When we hear a sound we wish to localize, we move our head in order to minimize the interaural differences, using our head as a sort of "pointer." Animals use movable pinnae for the same purpose. An informal observation of the household cat shows a reaction to an unfamiliar sound source as follows: first, it spreads its ears outward to receive the widest possible arc of sound, moving the head while maximizing interaural differences; and then, once the head is turned toward the source, the ears come up straight to fine-tune frontal localization.

Several studies have shown that allowing a listener to move her head can improve localization ability and the number of reversals (Wallach, 1940; Thurlow and Runge, 1967; Thurlow, Mangels, and Runge, 1967). Listeners apparently integrate some combination of the changes in ITD, IID, and movement of spectral notches and peaks that occur with head movement over time, and subsequently use this information to disambiguate, for instance, front imagery from rear imagery.

Consider a sound at right 30 degrees azimuth, which could potentially be confused with a source at right 150 degrees azimuth. A hypothetical listener would attempt to center this image by moving his head to the right, since the ITD and IID cues, in spite of their front–back ambiguity, still suggest that the source is somewhere to the right. If, after the listener moves his head rightward, the sound source becomes increasingly centered because interaural differences are minimized, it must be in the front. But if instead the differences become greater as the head is turned—i.e., the sound gets louder and arrives sooner at the right ear relative to the left—it must be to the rear.

Unlike natural spatial hearing, the integration of cues derived from head movement with stereo loudspeakers or headphones will provide false information for localizing a virtual source. With loudspeakers, a distortion of spatial imagery will occur when the head is turned to face a virtual sound source, since the processing to create the illusion depends on a known orientation of the listener. With headphones, the head movement has no effect on localization of the sound, a situation that does not correspond to actual circumstances.

Moving Sound Sources

Just as moving the head causes dynamic changes for a fixed source, a moving source will cause dynamic changes for a fixed head. While the visual system probably has a separate neural mechanism for detecting motion, evidence for an equivalent auditory "motion detector" has not been fully demonstrated. Cognitive cues are a large part of the sensation of motion; a monaural speaker, for example, can give the sensation of a speeding automobile on a racetrack, through the transmission of multiple, associative cues from experience. One of the main cues for a moving source is **Doppler shift** (the change in pitch associated with, e.g., a jet airliner passing overhead).

Under controlled experimental conditions, the **minimum audible movement angle** (abbreviated **MAMA**) is measured. Compared to the minimum audible angle for fixed sources, which is around 1 degree under optimal conditions, MAMAs can range up to 3 degrees under optimal conditions (Strybel, Manligas, and Perrott, 1992), but can increase as a function of movement velocity, location of movement, and sound source type. MAMAs are smallest for broadband sources; at movement velocities between 2.8–360 degrees/sec, it increases, since the minimum integration time for the auditory system for taking successive "snapshots" of the sound is exceeded (150–300 msec) (Grantham, 1986). Similarly, the MAMA will increase when the "storage time" of the auditory system is exceeded at around 5–10 sec (Middlebrooks and Green, 1991).

One independent variable that has been manipulated in several localization studies is the rise time of the amplitude envelope of a stimulus. Generally, after a certain threshold, the slower the rise time of the envelope, the worse localization becomes—especially in reverberant environments (Hartmann, 1983). This is because the inhibitory mechanism of the precedence effect diminishes under these circumstances and because the ITD envelope cue can no longer be analyzed within the region of a perceptual temporal integration "window." Within certain reverberant contexts, transient information can also be smeared over time at the point where the sound reaches the listener. This sensitivity of hearing to ITD transient information applies particularly to the spatial perception of moving sources; a sound trajectory defined by a source that has rapidly articulated transients is easier to localize than one with a slow amplitude envelope rise time.

SPECTRAL CUES PROVIDED BY THE PINNAE

Ambiguous ITD and IID Cues

Referring to Figure 2.8, note that a sound source at position **A** would theoretically produce an identical ITD and IID as a source at position **B** and similarly for sources at positions **x** and **y**. The statement is only theoretical because with a real person, ITDs and IIDs would never be *completely* identical unless, as in Figure 2.8, a spherical head is assumed, with effects of asymmetry, facial features, and the pinnae disregarded. But when ITD and IID cues are maximally similar between two locations, such as in the present example, a potential for confusion between the positions exists in the absence of a spatial cue other than ITD and IID.

In fact, identical values of ITD and IID can be calculated for a sound source in space anywhere on a conical surface extending out from the ear (see Figure 2.8). This is called the **cone of confusion** in the literature (see review by Mills, 1972). The cone of confusion has influenced the analysis and design of many localization studies in spite of the oversimplistic modeling of the head. The ability to disambiguate sources from front to back or from above and below in cases where ITD and IID would not supply this information has brought about hypotheses regarding the role of **spectral cues** and localization. In turn, the most significant locationally dependent effect on the spectrum of a sound source as it reaches the eardrums can be traced to the outer ears, or **pinnae.**

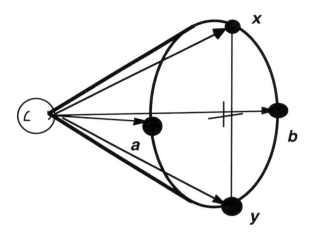

FIGURE 2.8. The cone of confusion. Identical values of IID and ITD can be calculated for a spherical model of a head at two opposite points anywhere on the surface of the cone (here, **a** and **b** show a front–back ambiguity, and **x** and **y** show an elevation ambiguity).

The Head-Related Transfer Function

The spectral filtering of a sound source before it reaches the ear drum that is caused primarily by the outer ear is termed the **head-related transfer function (HRTF)**. The **binaural HRTF** (terminology for referring to both left- and right-ear HRTFs) can be thought of as a frequency-dependent amplitude and time-delay differences that result primarily from the complex shaping of the pinnae.

The folds of the pinnae cause minute time delays within a range of 0–300 μsec (Batteau, 1968) that cause the spectral content at the eardrum to differ significantly from that of the sound source, if it were measured, for instance, by an omnidirectional microphone. The asymmetry of the pinnae's shape causes the spectral modification to be affected differently as a function of sound source position; together, the asymmetrical, complex construction of the outer ears causes a unique set of microtime delays, resonances, and diffractions that collectively translate into a unique HRTF for each sound source position (see Wright, Hebrank, and Wilson, 1974; Batteau, 1968; Shaw and Teranishi, 1968; Blauert, 1983). In other words, unique frequency-dependent amplitude and time differences are imposed on an incoming signal for a given source position. Like an "acoustic thumb print," the HRTF alters the spectrum and timing of the input signal in a location-dependent way that is recognized to a greater or lesser degree by a listener as a spatial cue.

The use of HRTF spectral shaping is typically featured as the key component of a 3-D sound system, from either direct measurement or modeling. This is based on the theory that the most accurate means to produce a spatial sound cue is to transform the spectrum of a sound source at the eardrum as closely as possible to the way it would be transformed under normal spatial hearing conditions. Within the literature, other terms equivalent to the term HRTF are used, such as Head Transfer Function (HTF), pinnae transform, outer ear transfer function (OETF), or directional transfer function (DTF).

Depending on the criteria set for a particular application, it may or may not be important to determine what features of the body in addition to the pinnae are taken into account for defining the HRTF. Some measurements account for only the outer ear and the head; others incorporate features of the body, such as the ear canal, shoulder, and torso, that also contribute somewhat less to the overall filtering characteristic (Gardner, 1973; Searle, et al., 1976; Genuit, 1984). This can be broken down into components that either vary with direction, or are constant with direction.

Figure 2.9 shows one type of breakdown of various HRTF components, in terms of directionally dependent and independent cues. The relative importance of the directional cues is arranged from bottom (most important) to top (Genuit, 1984; Gierlich, 1992). Gardner (1973) and Searle *et al.* (1976) cite a weak but nevertheless experimentally significant influence of the shoulders—the "shoulder bounce" cue. Genuit (1984) describes a directionally-dependent influence of the upper body and shoulders approximately between 100 Hz–2 kHz, particularly on the median plane. For instance, from the effect of the upper body alone, a sound from the front, relative to the back, would be boosted by about 4 dB from the front between 250–500 Hz and damped by about 2 dB between 800–1200 Hz. Gierlich (1992) summarizes the directional influence of the torso to be ± 3 dB, and ± 5 dB for the shoulder.

The **ear canal** (or "auditory canal") can be considered as an acoustical transmission line between the eardrum and the beginning of the outer ear. It is a tubular-shaped passageway with a mean length of 2.5 cm and a diameter of 7–8 mm, with variation in the size of the tube across its length (Blauert, 1983). Any acoustically resonated tube—e.g., a musical wind instrument—will effect a spectral modification to an incoming sound. For the ear canal, most measurements show a broad spectral peak similar to that shown in Figure 2.10. Luckily for most application purposes, this spectral modification has been verified in several experiments as essentially nondirectional (Blauert, 1983; Hammershøi, et al., 1992). It must be pointed out that measurement of the ear canal resonance is difficult to perform and can show a high degree of variance depending on the measurement technique used.

The **cavum conchae** is the largest resonant area of the pinnae, at the opening of the ear canal; the answer to whether or not it is directionally

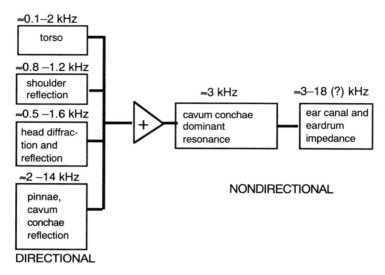

FIGURE 2.9. Gierlich's description of directional and nondirectional components of HRTFs (after Genuit, 1984; Gierlich, 1992). The range of frequencies most likely affected by each stage is indicated. The upper ranges of the ear canal resonance must usually be approximated by models.

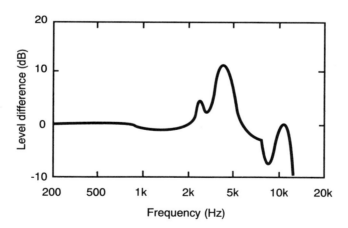

FIGURE 2.10. Ear canal resonance: the level difference between the eardrum and the entrance to the ear (adapted from Blauert, 1983). Actual measurements depend on the technique used, and there is usually a high degree of variability. For instance, a modeled ear canal (using a tube) shows spectral boosts of no more than 5 dB at 2.5, 7 and 14.5 kHz, and dips at 3.2, 10 and 18 kHz (Genuit, 1984).

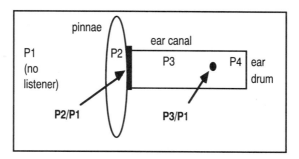

FIGURE 2.11. Measurement points for the HRTF (after Møller, 1992).

dependent depends on the frequency range measured and where the measurement within the cavum conchae is taken. Practically speaking, its directional aspects are included within the transfer function of the pinnae. A nondirectional "resonance" (Gierlich, 1992) is sometimes included as part of the ear canal resonance. Shaw and Teranishi (1968) explained differences in the magnitude of the HRTF based on location-specific standing waves within the area. Blauert (1983) reports experiments where he demonstrated the difference in the resonant effect of the cavum conchae on stimuli heard from the front compared to the back (0°–180°); a difference of around 5 dB occurred at 10 kHz.

Figure 2.11 shows the model given by Møller (1992) for showing the relationship between the measurement point for the HRTF and the ear canal. The HRTF is defined as the ratio between the sound pressure at the eardrum (P4) and at a point at the center of the head when the listener is absent (P1). HRTF measurements can be calculated either in terms of a point within the ear canal, P3/P1 (which includes both directional and nondirectional aspects of the HRTF) or at the entrance of a blocked ear canal, P2/P1, where only directional aspects are included. When this latter ratio is used for obtaining HRTFs from multiple directions, the nondirectional aspect P3 can be calculated once and then convolved with each measurement. Further information on HRTF measurement is covered in Chapter 4.

A Do-It-Yourself Experiment with Spectral Modification

The spectral modification of the pinnae can be thought of as equivalent to what a graphic equalizer does in one's stereo system: emphasizing some frequency bands, while attenuating other bands. This is notable from the standpoint that both timbre and spatial cues depend on the spectral modification of a sound source. Theile (1986) has addressed this via a descriptive model (see Figure 2.12) where the perception of timbre occurs at a later stage than that of spatial location; it is not the stimulation of the eardrum that determines timbre,

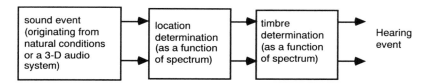

FIGURE 2.12. Simplification of Theile's model for the determination of timbre versus spatial location cues due to spectral content (adpated from Theile, 1986).

but rather the overall sense of hearing that "identifies the timbre and location of the sound source."

One can exaggerate spectral effects by altering how your own pinnae function with your hands. Try sitting close (about 9–12 inches) to a broadband sound source with high-frequency content, e.g., a TV speaker turned to a "blank channel," or the fan noise from a computer on a desk. Center the source at 0 degrees azimuth, close your eyes while directly facing the source, and listen for spectral (tone color) changes, comparing the "normal" condition with the pinnae unblocked (no hands) to each of the following conditions:

> **A.** With both hands flat and fingers together (as if saluting or giving a karate chop), create a "flap" in front of both ears to shadow the sound from the front, to focus incoming sound from the rear. Cup your fingers slightly over the top of the pinnae, so that the result is as if you had large "reversed pinnae" over the top of your ears.
> **B.** With your hands and fingers in the salute shape as in **A**, hold them flat out *under* the entrance to the ear canal, as if giving yourself a double karate chop on the head just below the ears. Experiment with angles less than 90 degrees between the hands and the top of the head.
> **C.** Flatten the pinnae back against your head with your fingers (make sure to fold as much of it as you can).
> **D.** Narrow the focus of the pinnae by pushing them out from your head with your fingers.

You should definitely hear the tone color change when comparing the "normal" and the "altered" conditions. Now repeat the exercise, but pay attention to **where** the sound is relative to the "normal" condition. Try it again while moving your head, as if you're attempting to center the source. You might hear the image position change, although it's difficult sometimes to make the "cognitive leap of faith" since you have prior knowledge as to where the source is. Some people notice spatial effects similar to the following when trying this exercise. Condition **A** can move the sound to the rear, when the head is held stationary; **B** can change the elevation; **C** spreads the sound image out, causing its location to be more diffuse; and **D**, besides making the sound louder, can also reduce the perceived distance. The latter situation can be seen with the movable pinnae of some mammals, who bring an auditory source into focus in this way.

HRTF Magnitude Characteristics

Figures 2.13–2.15 show examples of the HRTF from three persons measured in the same laboratory. These have been adjusted to show a usable signal–noise ratio of 60 dB; any spectral difference below this amount would probably be either imperceptible, or out of the overall range of the playback system. Note not only the spectral changes as a function of position, but also the differences between the three subjects. These differences are caused by the fact that pinnae vary in overall shape and size between individuals. While **individualized** or **personal HRTFs** are shown in Figures 2.13–2.15, it is important to note that most 3-D sound systems are incapable of using the HRTFs of the user; in most cases, **nonindividualized (generalized) HRTFs** are used. Nonindividualized HRTFs can be derived from individualized HRTF averages, although this can potentially cause the minima and maxima of the spectra to be diminished overall. Usually, a theoretical approach—physically or psychoacoustically based—is used for deriving nonindividualized HRTFs that are generalizable to a large segment of the population (see Chapter 4, "Calculation of Generalized HRTF Sets").

HRTF Phase Characteristics

There are also changes in interaural phase as a function of frequency that are caused by the HRTF. Phase delay as a function of frequency can be understood as follows: if a wideband sound source containing all audible frequencies were played instantaneously at the pinnae, some frequencies would arrive later than others at the eardrum. The delay can be measured in degrees by realizing that a phase delay of 360 degrees equals a delay of one cycle of the waveform; hence, a 1000-Hz wave delayed by 360 degrees has a time delay of .001 sec, a 100-Hz wave would be delayed .01 sec, etc. Figure 2.16 shows, for a frequency range of 500–4000 Hz, the unwrapped phase difference for a single person at 30-degree increments of azimuth from 0–150 degrees, at 0 degrees elevation. The bold lines show the conversion from phase delay to interaural time delays **td** at .25, .5, and 1 msec. Figure 2.17 shows the time difference more conveniently, in terms of the **group delay**; the time values on the y axis correspond to the time delays indicated by the bold bars in Figure 2.16.

The phase response at a single pinna is likely less critical for determination of localization than the interaural time difference. It is not clear to what degree the

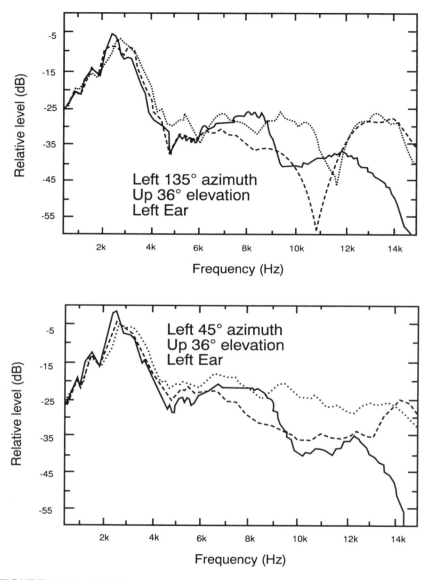

FIGURE 2.13. HRTFs on the "cone of confusion" for three people. *Data originally measured by Fred Wightman and Doris Kistler, University of Wisconsin—Madison.*

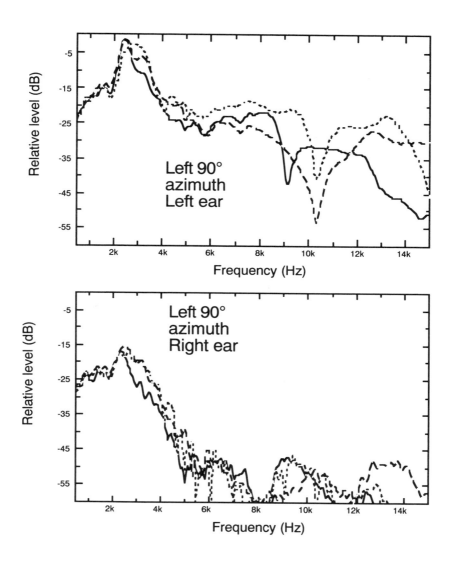

FIGURE 2.14. HRTFs, left and right ear (top and bottom), 90 degrees azimuth, for three different people. *Data originally measured by Fred Wightman and Doris Kistler, University of Wisconsin—Madison.*

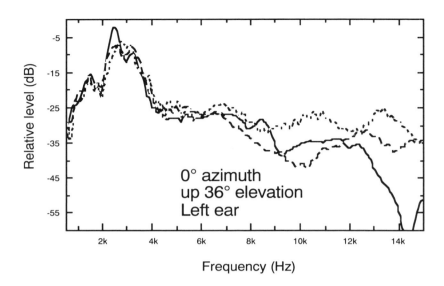

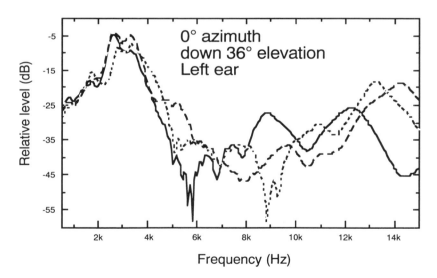

FIGURE 2.15. HRTFs, 0 degrees azimuth, up and down 36 degrees elevation, for three different people. *Data originally measured by Fred Wightman and Doris Kistler, University of Wisconsin—Madison.*

frequency-dependent delays are important, compared to the average interaural time delay across all frequencies. As with the transient envelope ITD cue, an overall estimate of the time delay is probably made based on the delay within each critical band.

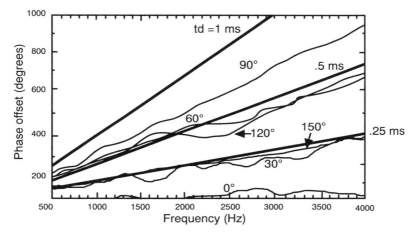

FIGURE 2.16. Unwrapped interaural phase difference for 0, 30, 60, 90, 120 and 150 degrees azimuth; one subject. Bold lines show interaural time differences of .25, .5 and 1 msec.

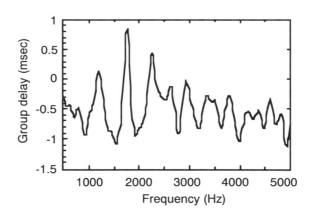

FIGURE 2.17. Group delay difference, left and right ear, left 60 degree azimuth.

Localization with HRTF Cues

Spectral Cues Provided by the HRTF

Within a 3-D sound system, inclusion of ITD and IID based on the spectral alteration of the HRTF is considered to be more accurate and realistic than the simple IID and ITD differences described previously in the discussion on lateralization. But this is probably true only in a qualitative sense, and not in terms of azimuth localization accuracy. A study by Middlebrooks (1992) examines the role of spectral cues as they contribute to accuracy of horizontal position. Only IID cues were used, with narrow bands of noise at high frequencies. He reasoned that if the HRTF spectrum contributed to IID localization independent of overall IID, then noises filtered into narrow bands should be localized with more error than full-spectrum noise that could "take advantage" of the HRTF. This turned out not to be true; azimuth location turned out to be just as accurate across both conditions.

The main role of HRTF cues from a psychoacoustic standpoint is thought to be the disambiguation of front from back for sources on the cone of confusion such as **X** and **Y** in Figure 2.8, and as an elevation cue for disambiguating up from down, e.g., the sound sources at **A** and **B** in Figure 2.8 (Roffler and Butler, 1968).

For example, Figure 2.18 shows the spectral difference in the HRTF at one ear of a single person, for two positions on the cone of confusion: 60 and 120 degrees azimuth (0 degrees elevation). The main difference occurs primarily in the upper frequencies, especially the broad peak at 5 kHz and the trough around 9 kHz. The relationship between spectral weighting and front–back confusions can

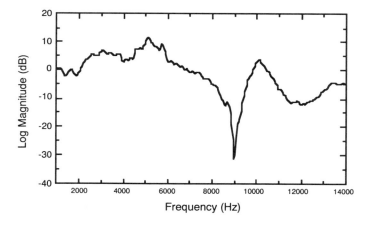

FIGURE 2.18. Difference in spectra between two front–back source locations on a cone of confusion: 60 and 120 degrees azimuth, 0 degrees elevation.

be seen in evaluations of reversals as a function of single frequencies. In one study, confusions were most pronounced in the frequency region where ITD "takes over" for IID, i.e., around 1.5 kHz (Mills, 1972). Stevens and Newman (1936) showed that sine waves below 3 kHz were judged as front or back on the order of chance, independent of actual location. This evidence, combined with the cone of confusion model and the fact that the binaural HRTF shows minimal IID below 400 Hz, suggests the role of spectral shaping at higher frequencies for front–back spectral cueing.

The directional effects of the pinnae are considered to be particularly important for vertical localization; several studies have shown that without the pinnae's effect on a broadband source, vertical localization is diminished (Gardner and Gardner, 1973; Oldfield and Parker, 1984b). To examine the HRTF spectral cues for elevation, some studies have isolated the role of ITD and IID by examining the special case of sources along the **median plane** of the listener; i.e., the plane that intersects the left and right sides of the head, at 0 and 180 degrees azimuth. Notice that this includes those positions where, besides IID and ITD, the spectral difference would be minimized between the HRTFs to the degree that the left and right pinnae are identical. The role of the pinnae is also evaluated by comparing judgments made under normal conditions to a condition where the pinnae are bypassed or occluded (Musicant and Butler, 1984; Oldfield and Parker, 1984b). These studies almost always show that the pinnae improve vertical localization. Other experiments where one ear was blocked support the fact that spectral cues work **monaurally**; i.e., the **spectral difference** between the HRTFs is considered to be less important than the overall spectral modification at a particular ear (Middlebrooks and Green, 1991), although there is some evidence for binaural pinnae cues (Searle, et al., 1975).

Theories surrounding the role of HRTF spectral cues involve **boosted bands** (Blauert, 1969), **covert peaks** (Butler, 1987), and spectral **troughs**, or notches (Shaw and Teranishi, 1968; Bloom, 1977; Watkins, 1978; Rodgers, 1981). Butler and Belendiuk (1967), Hebrank and Wright (1974) and Rodgers (1981) all suggest that a major cue for elevation involves movement of spectral troughs and/or peaks. Changes in the HRTF spectrum are thought to be effective as spatial cues, to the degree that certain features of the magnitude function change as a function of source and listener orientation. For example, the movement in the center frequency of the two primary spectral troughs apparent in Figure 2.19 could contribute to the disambiguation of front–back source positions on the cone of confusion, especially with head movement. Figure 2.20 shows a similar spectral trough that is thought to be important for elevation perception. But it is difficult without extensive psychoacoustic evaluation to ascertain how importantly these changes function as spatial cues. In one study, a comparison of spectral peaks and notches at 1 kHz and 8 kHz suggested that the cues provided by HRTF peaks were more salient than the troughs (Moore, Oldfield, and Dooley, 1989). Also, it is unclear if localization

FIGURE 2.19. HRTF magnitude changes on a KEMAR mannequin head as a function of azimuth at a single elevation *(courtesy William Martens)*. Note the two major spectral troughs that change center frequency as a function of azimuth.

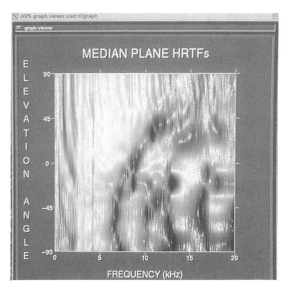

FIGURE 2.20. HRTF magnitude changes on a KEMAR mannequin head as a function of elevation at 0 degrees azimuth *(courtesy William Martens)*. Note the spectral trough around 8–10 kHz that migrates with elevation changes.

cues are derived from a particular spectral feature such as a "band" or a "trough," or from the overall spectral shape.

Finally, it is generally considered that a sound source contain broadband spectral energy in order to be affected by the particular cue. For instance, most studies have determined that vertical localization requires significant energy above 7 kHz (Roffler and Butler, 1968). Noble (1987) determined from his own studies and analysis of other studies that around 25 percent of the population exhibits little or no sensitivity to elevation judgments based strictly on spectral cues (i.e., without cognitive cues such as familiarity).

Spectral Band Sensitivity

The evidence does suggest that spectral cues can influence the perception of direction independent of the location of a sound source. Blauert (1969) used third-octave–filtered bands of noise across frequency to determine the directional bias associated with the spectral band, in terms of three categories—above, in front, or behind. Table 2.1 gives the approximate regions where it was determined that "directional bands" existed. The relationship between directional bands and the spectral peaks of the HRTF led to the formation of a theory of "boosted bands" as a spatial cue.

Middlebrooks (1992) used narrowband noise stimuli at high frequencies: 6 kHz, 8 kHz, 10 kHz, and 12 kHz. Note that the spectral content of a broadband noise would be affected by all frequency regions of the HRTF, but with narrowband noise, the "spectral shaping" characteristics only affect the narrow passband used in the experiment. An applications-relevant result of Middlebrooks's study was the biases of judgments for elevation up–down, and front or back, as a

Perceived location	Center frequency kHz	Bandwidth kHz
overhead	8	4
forward (band #1)	0.4	0.2
forward (band #2)	4	3
rear (band #1)	1	1
rear (band #2)	12	4

TABLE 2.1. Center frequencies and frequency bandwidths for "directional bands" (Blauert, 1983). A broadband noise that is filtered to contain primarily the frequencies shown in the second and third columns will be heard from the direction indicated in the first column.

Fc	Overall results		Tallest subject		Shortest subject	
	u/d	f/b	u/d	f/b	u/d	f/b
6 kHz	U	F	U	F	U	F
8 kHz	D	F & B	D	F	U	B
10 kHz	D	B	level-D	F	D	B
12 kHz	level-D	F & B	level-D	F	D	B

TABLE 2.2. Results for narrowband filtered noise, adapted from Middlebrooks (1992). The noise stimuli used had four different center frequencies (left column), and target stimuli were from a variety of azimuth and elevations. In spite of the target location, subjects exhibited overall biases of direction, both up-down (**u/d**) relative to eye level, or front-back (**f/b**) relative to the interaural axis. The overall results for five subjects are compared to the results for the tallest and shortest subjects. **U** = up; **D** = down; **F** = front; **B** = back.

function of the center frequency of the narrowband noise. These localization biases were independent of the target direction. There were also significant individual differences, probably due to the fact that the five subjects were chosen to span a range of heights; Table 2.2 compares the results of the shortest and tallest people to the overall results for the five subjects evaluated. The working assumption was that the size of the pinnae, and therefore the pinnae's primary spectral features, are related to the overall height of a person (see also Middlebrooks and Green, 1991).

Taken together, the results of Blauert (1969) and Middlebrooks (1992) show that directional biases can result from stimuli that mimic particular spectral modifications of the pinnae. However, the effect of these spectral modifications will depend on the particular individual. It would be unlikely because of these differences that a device that merely emphasized particular frequency bands for a specific effect would be practical for inclusion within a basic 3-D sound system (e.g., Myers' (1989) US Patent, described in Chapter 6). Rather, a full implementation of spectral cues provided by the HRTF is desirable. The effect of the spectral modification of the entire pinnae is examined ahead, i.e., the net effect of the overall HRTF.

Localization of Actual Sound Sources with HRTF Cues

Before it was easily possible to synthesize HRTF cues electronically with 3-D sound systems, psychoacousticians studied localization ability by placing loudspeakers at different locations and then asking the subjects to report the apparent location of the source in a categorical manner. In most cases, an unblindfolded subject would choose which of several visible speakers was producing the sound. This type of experimental design biases results since the options available for where the subject hears the sound in auditory space are limited to the choices given by the experimenter. In addition, the subject is reporting the apparent location of the sound source, as opposed to where the spatial image is within his or her mind, an important distinction clarified mainly by Blauert (1983). Modern studies of localization allow the subject to report the location of sounds in a less biased manner. For example, the subject is typically blindfolded within an anechoic chamber and allowed to point at where the spatial image seems to be. Verbal reporting methods have also been used with success. Nevertheless, it is a fact that no psychoacoustic response paradigm can ever be completely "pure" of bias. Psychoacousticians therefore attempt to minimize bias to an acceptable degree within their experimental design.

Some trends seen in these earlier studies have yet to be refuted, in spite of newer, less biased methodologies. Fisher and Freedman (1968) placed loudspeakers with numerical identification marks at equal-spaced positions on the azimuth surrounding the subject and asked which speaker was sounding. This study shows two important results: (1) with head movement, categorical estimations are accurate, both with and without pinnae; and (2) with head movement restricted, either individualized or nonindividualized pinnae cues are sufficient for accurate categorical localization on the azimuthal plane; but when the pinnae are bypassed using tubes, categorical estimations are diminished.

Oldfield and Parker (1984a, b; 1986) used a method where the subject pointed to ("shot") the location of the apparent sound source location with a light gun; the data were analyzed from videotapes that allowed extraction of all three dimensions. (They were careful to evaluate their results in light of the error inherent in the response paradigm, which turned out not to affect the overall outcome of the results.) Their results are shown in Figure 2.21 for normal pinnae conditions, and in Figure 2.22 for occluded pinnae, i.e., a condition where the intricate folds of the outer ear were covered with a putty substance. Overall, their data for normal hearing conditions show localization absolute error has a mean value of approximately 9 degrees for azimuth and 12 degrees for elevation. When the pinnae's filtering effect is removed, azimuth absolute error increases to 11.9 degrees, but elevation ability degrades substantially, to 21.9 degrees. Figures 2.21–2.22 also show that localization ability tends to be variable, depending on the source direction; localization was worst for the upper rear hemisphere for normal conditions. Oldfield and Parker (1986) also investigated one-ear listening with no pinnae occlusion: the results showed that elevation judgments were hardly affected, but azimuth accuracy decreased substantially.

Figure 2.23 shows the results obtained by Middlebrooks (1992) with broadband noise, for a single person. This subject was blindfolded and used a pointing technique, where the results were calculated from a Polhemus head tracker when the subject indicated she or he had faced the source. The solid circles represent the target locations (i.e., the speaker locations), and the open

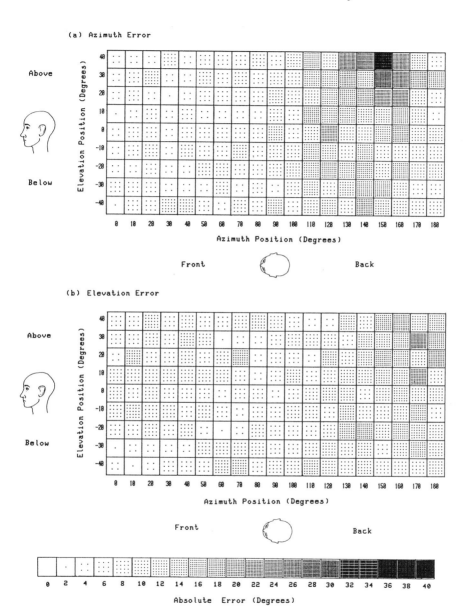

FIGURE 2.21. Oldfield and Parker (1984a): localization judgments of white noise, under free field conditions (mean values for eight subjects). *Courtesy of Simon Oldfield and Simon Parker; reprinted by permission of Pion Limited, London, from* Perception, *vol. 13, 1984, pages 581-600.*

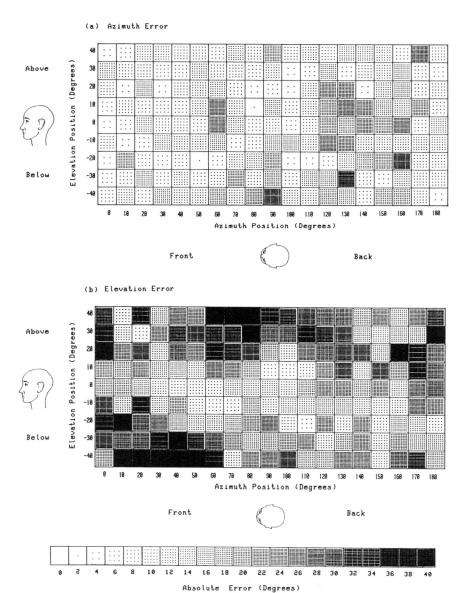

FIGURE 2.22. Oldfield and Parker (1984b): localization judgments of white noise, under free field conditions, but with pinnae folds occluded. *Courtesy of Simon Oldfield and Simon Parker; reprinted by permission of Pion Limited, London, from* Perception, *vol. 13, 1984, pages 601-617.*

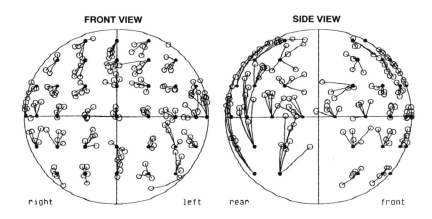

FIGURE 2.23. Localization accuracy for broadband noise (Middlebrooks, 1992). *Courtesy of John Middlebrooks; by permission of the American Institute of Physics.*

circles represent the judgment locations for the particular target. Note that errors are greatest in the rear, similar to the results of Oldfield and Parker (1984a) seen in Figure 2.21.

Localization of Azimuth and Elevation of Virtual Sources

The advent of 3-D sound hardware and an understanding of filtering techniques for artificially producing sounds in space were probably first anticipated by Bauer (1961) and Batteau (1966). Bloom (1977) and Watkins (1978) used analog filters designed to produce the essential elevation resonances of the pinnae within localization experiments. The technique was also used relatively early on by Blauert and Laws (1973). The use of digital filtering techniques and computers in descriptive localization studies is first seen in papers by Rodgers (1981) and Morimoto and Ando (1982). Wightman and Kistler (1989a, b) published one of the earlier full-scale descriptive studies of a subject's ability to localize under free-field conditions, compared to his or her ability to localize headphone virtual sources, with stimuli produced using 3-D sound techniques. They used a verbal response technique in their studies, where a subject was trained to imagine a vector from the center of his or her head to the center of the virtual sound source; the subject would then report, e.g., "left 0 degrees, up 30 degrees," after being trained for a period of time in the technique. Figure 2.24 shows the results from one of the subjects typical of "good localizers"; the points represent "centroids," which can be thought of as the mean direction of a set of judgment vectors. The

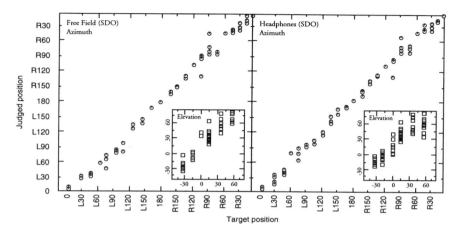

FIGURE 2.24. Data from Wightman and Kistler (1989b) for a single subject listening in free space (left) and under 3-D audio conditions (right). *Courtesy of Fred Wightman and Doris Kistler, University of Wisconsin—Madison; by permission of the American Institute of Physics.*

target positions are shown on the *x* axis, and judgments on the *y* axis; perfect localization would occur across a diagonal of the graph. A close agreement between judged and target positions can be seen under both conditions.

Nonindividualized HRTFs

A significant problem for the implementation of 3-D sound systems is the fact that spectral features of HRTFs differ among individuals, as shown previously in Figures 2.13–2.15. Hence, it makes sense to determine how localization of virtual sound sources can be degraded when listening through another set of pinnae. Fisher and Freedman (1968) showed a significant decrease in azimuth localization accuracy when listening through artificial pinnae versus the subject's own pinnae. Wenzel *et al.* (1988) conducted a study that showed that a person who was a relatively good localizer in elevation judgments would degrade his or her ability when listening through the pinnae of a person who was inherently worse in free-field elevation judgments. While the converse question—could the bad localizer improve listening through a good localizer's ears—has not been fully evaluated, the idea is nevertheless intriguing that people could improve their performance by listening through pinnae with "supernormal" auditory localization cues (Butler and Belendiuk, 1977; Shinn-Cunningham, et al., 1994).

The idea of using the pinnae of a "good localizer" was taken into a study of localization accuracy where several listeners gave judgments of localization through pinnae that were not their own—nonindividualized HRTFs. Figure 2.25 shows the results of Wenzel *et al.* (1993), in terms of the average angle of error

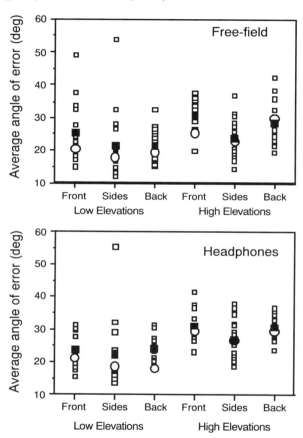

FIGURE 2.25. Individual and mean of 16 subjects' average angle of localization error for various spatial regions (see text for definition of the regions). From Wenzel *et al.* (1993). The top graph shows free-field condition responses (i.e., using the subject's own pinnae); the bottom graph shows responses for virtual sources synthesized with a single "nonindividualized" HRTF set of a good localizer (the pinnae of the subject whose data was shown previously in Figure 2.24). Data of Wightman and Kistler (1989b) are shown for comparison; in that study, the virtual sources were synthesized using the pinnae of each individual subject. *Courtesy Elizabeth Wenzel; used by permission of the American Institute of Physics.*

for various regions about the listener. The regions are defined as follows: front (left 45 to right 45 degrees); sides (left 60 to left 120 degrees, and right 60 to right 120 degrees); back (left 135 to right 135 degrees); up (up 18 to up 54 degrees); and down (down 36 to down 0 degrees). Overall, the error is similar between the virtual and headphone conditions, especially the overall trend of the data—e.g., judgments of higher elevations are worse than lower elevations (compare to Oldfield and Parker's data in Figure 2.21). An overall degradation in elevation localization was seen within a subgroup of listeners, and an increase in the number of front–back and up–down reversals was observed, compared to free-field conditions where the subject's own pinnae were used.

One aspect of localization that becomes obvious especially with nonindividualized HRTFs is the variation of performance between individuals. To illustrate this, the results obtained for nonindividualized HRTF-filtered speech are shown in Figures 2.26–2.28 (Begault and Wenzel, 1993). Compared to the broadband noise used in most experiments, speech is relatively low pass: the majority of energy is below 7 kHz. The speech stimuli were all presented at 0 degrees elevation, from 12 azimuths corresponding to "clock positions"—i.e., left and right 30–150 degrees, and 0 and 180 degrees. Figure 2.26 shows results for subjects who localized relatively accurately; Figure 2.27 shows results for subjects who "pulled" judgments toward the plane intersecting left and right 90 degrees. Figure 2.28 shows responses of an additional subject who pulled his judgments in a similar way, but toward the plane intersecting 0 and 180 degrees.

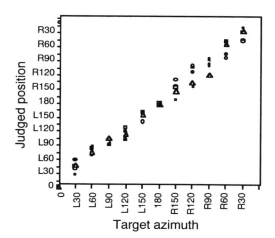

FIGURE 2.26. Data for HRTF-filtered speech stimuli, showing data for six subjects who made relatively accurate judgments (Begault and Wenzel, 1993). *Reproduced with permission from* Human Factors, *Vol. 35, No. 2, 1993. Copyright 1993 by the Human Factors and Ergonomics Society. All rights reserved.*

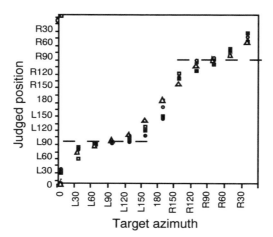

FIGURE 2.27. Data for speech stimuli, showing data for four subjects whose judgments "pulled" toward left and right 90 degrees (Begault and Wenzel, 1993). *Reproduced with permission from* Human Factors, *Vol. 35, No. 2, 1993. Copyright 1993 by the Human Factors and Ergonomics Society. All rights reserved.*

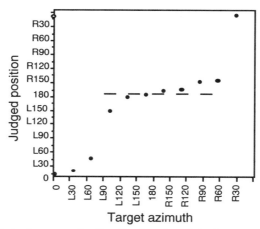

FIGURE 2.28. Data for speech stimuli, showing data for one subject whose judgments "pulled" toward the median plane: 0 and 180 degrees (Begault and Wenzel, 1993). *Reproduced with permission from* Human Factors, *Vol. 35, No. 2, 1993. Copyright 1993 by the Human Factors and Ergonomics Society. All rights reserved.*

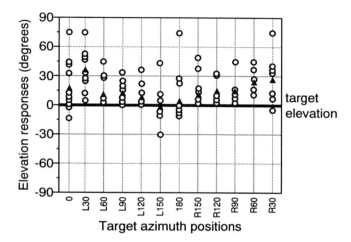

FIGURE 2.29. Elevation judgments for speech stimuli; mean judgment centroid for each subject is shown as the open circle; triangles show the mean across 11 subjects. *Reproduced with permission from* Human Factors, Vol. 35, No. 2, 1993. *Copyright 1993 by the Human Factors and Ergonomics Society. All rights reserved.*

Figure 2.29 shows the elevation responses for each target azimuth; recall that the target elevation was always eye level at 0 degrees. For reasons that are not entirely clear, the data show an overall bias toward elevated judgments (an average of up 17 degrees). But while this is true for the speech stimuli used here, the equivalent study by Wenzel *et al.* (1993) using noise bursts shows no particular elevation bias.

Martens (1991) examined the apparent location (and particularly the elevation) of HRTF-filtered speech stimuli by using short segments of speech (e.g., /i/, /ae/, /a/, and /u/). These syllables varied in terms of their perceived timbral "brightness" as a function of spectral content. The subjects in the experiment listened to this stimuli as processed through nonindividualized HRTFs (obtained from a KEMAR mannequin head; see Chapter 4) at various elevation positions at 90 degrees azimuth. Elevation judgments were found to be accurate in this experiment, but the front–back position was influenced by brightness; specifically, stimuli with energy concentrated at higher spectral frequency regions tended to be heard to the front, and stimuli concentrated at lower spectral frequency regions were heard to the rear. Surprisingly, when the stimuli were low-pass–filtered at 5 kHz, the judgments of elevation remained accurate. This contrasts the notion that elevation perception is necessarily based on higher-frequency, monaural spectral features of the HRTF (Roffler and Butler, 1968). This led Martens (1991) to investigate the interaction between binaural versus monaural presentation of phase and magnitude features of the HRTF as a function of different elevations. It turned out that elevation perception was still accurate when one of these features was removed, but any one of these cues was in itself insufficient to HRTF to create accurate elevation simulations— including monaural magnitude changes in the HRTF.

To emphasize the difference in localization among listeners, individual results from 11 subjects are shown in Figure 2.30 from the Begault and Wenzel (1993) study. These data correspond to the judgment centroids seen in Figures 2.26– 2.28. The "ideal responses" for azimuth that would correspond exactly to the target positions are given as a reference in the upper left corner. Distance estimates are also shown; the inner circle represents a distance of 4 inches, which subjects used as a reference for the edge of the head.

Reversals

It is important to realize that the previous descriptive localization judgments have been "reversal corrected"; e.g., if a judgment resulted in less error when "folded" across the interaural axis, it is analyzed in its reversal-corrected form. For example, if the target position was right 30 degrees azimuth and the judgment was 130 degrees, the judgment is recoded as 50 degrees (see Figure 2.31). The approach used by most experimenters is to report on the number of reversals separately; this keeps the descriptive location plots from being artificially inflated, since reversals seem to be a function of different factors than those responsible for localization error.

Most data show a trend for more front–back reversals than back–front reversals; for instance, in the study of speech stimuli by Begault and Wenzel (1993), the ratio of front–back reversals was 3:1 (about 50%, but only 15% for back–front reversals, with speech stimuli and nonindividualized HRTFs—see Figure 2.32). This is sometimes explained (although not proven) as originating from a primitive survival mechanism that assumes that "if you can hear it but can't see it, it must be behind you!"

Table 2.3 summarizes reversals under various conditions among several experiments. Figure 2.33 shows the results obtained by Wenzel et al. (1993) for nonindividualized HRTFs, compared to the results for the individualized HRTFs used by Wightman and Kistler (1989b). Note the higher reversal rates found for headphone versus free-field stimuli. The reversal rates for Wightman and Kistler's data are lower under both free-field and 3-D sound conditions, probably since their subjects listened through individualized HRTFs.

Reversals are very common in both free-field and virtual acoustic simulations of sound. For instance, Stevens and Newman (1936) demonstrated early on that relatively low-frequency sine waves are almost ambiguous as to their front–back location for subjects listening to loudspeaker sound sources. Several reasons are given by various researchers. The most plausible reason is that turning the head can disambiguate a source's front–back location over time (see earlier discussion

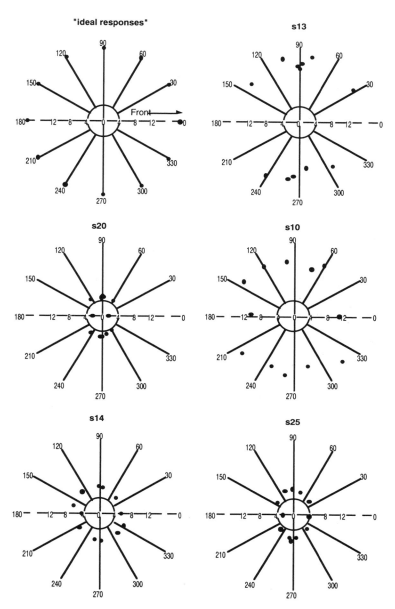

FIGURE 2.30. Azimuth and distance judgments for speech stimuli (Begault and Wenzel, 1993: see text). Upper left picture shows "ideal responses." *Reproduced with permission from* Human Factors, Vol. 35, No. 2, 1993. *Copyright 1993 by the Human Factors and Ergonomics Society. All rights reserved.*

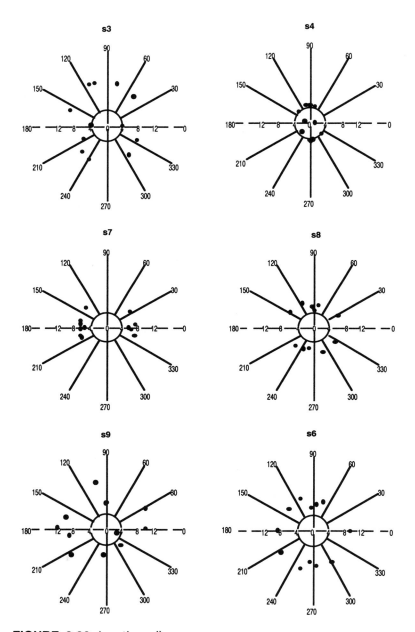

FIGURE 2.30 (continued).

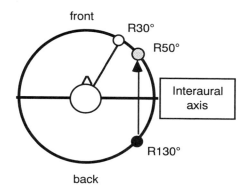

FIGURE 2.31. Recoding of a reversed judgment in a localization experiment (see text).

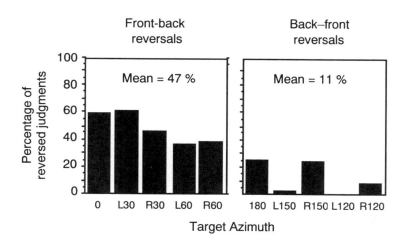

FIGURE 2.32. Reversed judgments for speech stimuli, nonindividualized HRTFs (from Begault and Wenzel, 1993). *Reproduced with permission from Human Factors, Vol. 35, No. 2, 1993. Copyright 1993 by the Human Factors and Ergonomics Society. All rights reserved.*

on head movement and integration). Another reason is that listeners are able to utilize small differences in their own HRTF characteristics to distinguish front from back (Weinrich, 1982). Spectral content is probably also important; sound sources tend to be reversed more often when the spectral bandwidth is relatively narrow (Burger, 1958; Blauert, 1969).

Study	Free-field reversals	Virtual source reversals	Ratio B–F to F–B reversals
Burger (1958); one-octave noise bands	20%		
Laws (1974); 0° target only	0	35%	n/a
Oldfield and Parker (1984a, b)	3.4% (12.5% monaural listening; 26% pinnae occluded)	n/a	
Wightman and Kistler (1989b); user's own HRTF	6%	11%	2:1
Makous and Middlebrooks (1990)	6%	n/a	
Asano *et al.* (1990); 2 subjects	n/a	**subject 1**: 1.9% with own HRTF: 0.9% with other subject's HRTF **subject 2**: 7.4% with own HRTF: 14.8% with other subject's HRTF	
Wenzel *et al.* (1993)	19%	31%	free-field: 8.5:1 virtual: 4.0:1
Begault and Wenzel (1993); speech	n/a	37%	3:1

TABLE 2.3. Overall percentage of reversals of sound source positions under various conditions. Broadband noise was the stimulus in all studies except Begault and Wenzel (1993) and Burger (1958).

Knowledge of the interaction of a sound source with the acoustical environment might provide further cues to front–back discrimination; for instance, if a reverberant space were to the rear and a solid wall was directly in

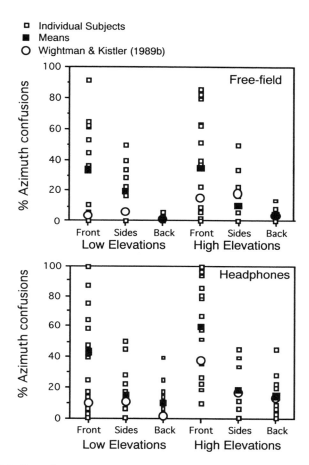

FIGURE 2.33. Data from Wenzel *et al.* (1993) showing regional tendencies for reversals. *Courtesy Elizabeth Wenzel; used by permission of the American Institute of Physics.*

front between the listener and the sound source, knowledge of the environment might help disambiguate the location of the source within it (Begault, 1992a). Another reason given by a manufacturer of mannequin heads is a bit suspect: the company claims that reversals do not occur for younger persons or nontechnical persons who are not aware of the publicity surrounding the reversals that were frequently reported with the first binaural recordings of the 1970s (Steickart, 1988).

One of the most plausible explanations of reversals has to do with learning to **categorize** sounds on the basis of the resultant spectral modification. The author can report informally that, upon listening for the first time through stimuli processed by individualized as opposed to nonindividualized HRTFs, it was much easier to decide spontaneously which one "should" appear to the front and which one to the back, perhaps supporting Weinrich's (1982) conclusion about the importance of individual pinnae differences. The finding by Martens (1991) that "brighter"-sounding speech appeared forward of the interaural axis and that "darker"-sounding speech appeared to the rear, and Blauert's (1969) "boosted bands" support a type of categorization based on overall estimates of spectra. No formal research has been done in the area of learning and reinforcement, i.e., where a subject listens through nonindividualized HRTFs and is "rewarded" via positive feedback when she learns to associate spectral cues with front and back. Perhaps, as some have suggested, one can adapt to the nonindividualistic HRTFs contained within a 3-D sound system: ". . . listeners adapt to overall frequency response errors, and the spatial image gradually becomes quite realistic" (Griesinger, 1989).

CHAPTER THREE

Overview of Spatial Hearing

Part II: Sound Source Distance and Environmental Context

3-D SOUND, DISTANCE, AND REVERBERATION

The inclusion of distance and environmental context effects within a 3-D audio system is almost imperative for maintaining a sense of realism among virtual acoustic objects. In fact, the same physical cue— reverberation—is key for the effective simulation of distance and the perception of an environmental context.

Even with nonspatial cues, one can tell something about the distance of different sounds and their context; for instance, you can experience distance listening with one ear, and you can tell sometimes in a telephone conversation where a person is calling from, based on the type of background noise. But the inclusion of 3-D sound techniques and spatial reverberation cues can greatly increase the immersiveness of a simulation, as well as the overall quality and the available nuance in representing different environmental contexts.

Further investigations of distance and environmental context effects have been facilitated by the development of more computationally powerful 3-D sound systems, which facilitates a larger range of experimental contexts that match natural spatial hearing. These areas are in need of further research, since most virtual reality and multimedia systems currently offer rather crude or nonexistent reverberation and distance simulations. But with the advent of computer-assisted room modeling algorithms and binaural measurements of actual halls, the spatial aspects of reverberation are now receiving increased attention. This latter area is known as **auralization**, and its commercial potential should help motivate a drive toward truly veridical room simulations. Potentially, such systems will allow enough detail to reveal how well, for instance, the noise from a city street is absorbed by the wall materials in a new building before it is built—to *hear* what the new building will sound like, from within any room.

Distance and environmental context perception involve a process of integrating multiple cues, including loudness, spectral content, reverberation content, and cognitive familiarity. Unfortunately, psychoacousticians must minimize the number of variables present in a particular study so as to obtain consistent, predictable results. For instance, reverberation is usually not included in distance perception studies, thereby giving subjects an incomplete description of the way a stimulus would be heard under natural spatial hearing conditions. Estimation of the environmental context is not possible, except with reference to anechoic conditions; and the estimation of distance is usually worse than in experiments with "optimal" reverberation conditions (Gardner, 1969; Mershon and King, 1975; Sheeline, 1982). The reader should therefore take into account that any one of the cues discussed below can be rendered ineffective by the summed result of other potential cues.

DISTANCE CUES

Auditory distance studies require classification according to two types of experimental methods in order to assess their application to a particular situation. The first methodological difference is whether an **absolute** or **relative** sense of distance is being evaluated. Absolute distance perception refers to a listener's ability to estimate the distance of a sound source upon initial exposure, without benefit of the cognitive familiarity. Relative distance perception includes the benefits gained from listening to the source at different distances over time, perhaps within an environmental context. (Another form of relative estimates in an experiment is when listeners adjust a sound to be "twice as distant" as another sound, irrespective of measurement units such as inches or meters).

The second methodological difference has to do with whether a listener is asked to estimate the apparent distance of the **virtual sound image** — e.g., "the sound seems to be 20 inches away from the outside of my head" — or to estimate the distance of the actual **sound source**. As an example of the latter case, a subject might be asked to choose from five loudspeakers, labeled 1–5. This of course disallows the possibility of the sound coming from any other distance than the five speakers chosen by the experimenter. Furthermore, the **ventriloquism effect** comes into play, whereby the apparent position of an virtual sound image is influenced by the presence of a correlated visual object. For 3-D sound system applications, the most relevant psychoacoustic studies are those where subjects give distance estimates of the virtual sound image, without visual cues; since most applications involve extended use, both relative and absolute judgments are useful. What would be particularly relevant to virtual reality applications — but, because of newness of the technology, has yet to be systematically studied — are studies of the interaction between visual objects and aural distance cues in a virtual world.

Intensity, Loudness Cues

In the absence of other acoustic cues, the **intensity** of a sound source (and its interpretation as **loudness**) is the primary distance cue used by a listener. Coleman (1963) stated in a review of cues for distance perception that, "It seems a truism to state that amplitude, or pressure, of the sound wave is a cue in auditory depth perception by virtue of attenuation of sound with distance." From one perspective, auditory distance is learned from a lifetime of visual-aural observations, correlating the physical displacement of sound sources with corresponding increases or reductions in intensity. This may be the primary means we use for many everyday survival tasks, for instance, knowing when to step out of the way of an automobile coming from behind us. Intensity adjustments are of course ubiquitous in the world of audio. A recording engineer, given the assignment to distribute a number of sound sources on different tracks of a multitrack tape to different apparent distances from a listener, would most probably accomplish the task intuitively by adjusting the volume of each track. Terminology taken from the visual world, such as "foreground" and "background," are usually used in these contexts; most audio professionals don't get more specific than this verbally, although in practice the distance-intensity relationships of a multitrack recording mixdown can be quite intricate.

As a cue to distance, loudness or intensity probably plays a more important role with unfamiliar sounds than with familiar sounds. In other words, exposure to a particular sound source at different distances allows an integration of multiple cues over time for distance perception; but without this exposure, cues other than loudness fall out of the equation. A good example is a comparison of the experience of listening to sounds just before going to sleep, in an unfamiliar versus a familiar environmental context. In the familiar environment, say an apartment in a city, you know the distance of the sound of the local bus passing by and of the ticking clock in the kitchen. Although the bus is louder than the clock, familiarity allows distance estimations that would be reversed if intensity were the only cue. But when you camp outdoors in an unfamiliar environment, the distance percepts of different animal noises or of rushing water would probably follow an intensity scale.

What is the relationship between sound source distance and intensity change at a listener? Under anechoic conditions, one can use the **inverse square law** to predict sound intensity reduction with increasing distance. Given a reference intensity and distance, an omnidirectional sound source's intensity will fall almost exactly 6 dB for each subsequent doubling of distance from a source. Figure 3.1 shows the inverse square law for two spherical distances from an omnidirectional source. If the sound source was not omnidirectional but instead a **line source**, such as a freeway, then the intensity reduction is commonly

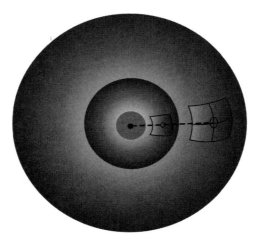

FIGURE 3.1. The inverse square law. An omnidirectional sound source (shown as the dotted point in the center) within an anechoic environment is assumed. The area described on the inner sphere as a square represents a reference intensity of one watt per square meter. Because the surface area of a sphere is a function of the square of the radius, the sound energy would be spread over four times the area on the outer sphere if the radius were twice that of the inner sphere. The intensity would now be 1/4 watt per square meter, or about 6 dB less.

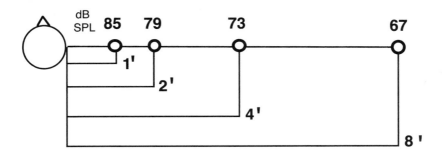

FIGURE 3.2. Reduction in intensity for virtual sources, using the assumptions of the inverse square law. Intensity falls by 6 dB for each doubling of distance.

adjusted to be 3 dB for doubling of distance. This latter adjustment is found in noise-control applications but almost nowhere in virtual acoustic simulations; almost always, an omnidirectional point source is assumed.

In Figure 3.2, four identical sound sources in an anechoic chamber at 1, 2, 4, and 8 feet from a listener's right ear are shown. Given the reference intensity of 85 dB SPL for the one-foot-distant sound source, the intensity would follow incremental reductions of 6 dB for each successive position of doubled distance, according to the inverse square law. Most 3-D sound systems use such a model for calculating the reduction in intensity, without reference to absolute sound pressure level. Usually, the maximum digital value (e.g., +32767 in a signed integer system) is used for the "most proximate" virtual sound source, and then a coefficient is used for scaling intensity in relation to distance.

In a 3-D sound system using the inverse square law, one could specify several sound sources at progressively doubled distances. If the most proximate sound source was known to be 85 dB SPL at the eardrum, and the noise floor of the audio system was at 40 dB SPL, then hypothetically it would be possible to place the sounds at eight different relative positions, each doubled in distance from the other. Along the same line of thinking, if the source at 85 dB SPL was perceived at 12 inches from the head, as in Figure 3.2, the most distant source at 43 dB SPL would be perceived at a point 128 feet away.

A theoretical problem with the inverse square law as an effective cue to distance lies in the fact that perceptual scales of loudness are unaccounted for. Intensity, when measured in dB, expresses the ratio of a sound source's intensity to a reference level, but loudness is the *perceived* magnitude of intensity, as pointed out in Chapter 1. Given that a perceptual scale is desired in a 3-D sound system, and that loudness is the only available cue for distance involved, a mapping where the relative estimation of doubled distance follows "half-loudness" rather than "half-intensity" seems preferable; as will be seen, the two scales are different. Extending this from relative to absolute measurements, a loudness scale can be used to adjust intensity for scaling the apparent distance of virtual sound source images.

Judgments of half- or doubled loudness have been shown to be more closely related to the **sone scale** than to the inverse square law (Stevens, 1955). Sones are a unit that relate equal loudness contours (**isophons**) for a given frequency to loudness *scales*. The scales were derived from studies where subjects were asked to adjust sine waves to be twice as loud; double the number of sones, and you've doubled the loudness. A symphony orchestra can range in loudness from 40–100 dB SPL; this ranges from about 1–50 sones, or between five and six doublings of loudness (Pierce, 1983). It turns out that, for "most sounds" (sine waves lying on relatively "flat" isophons between 400–5000 Hz and 40–100 dB SPL), an increase of 10 dB is roughly equivalent to a doubling of loudness.

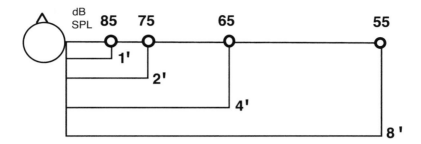

FIGURE 3.3. Reduction of intensity, using a loudness scale based on sones. Compare the intensity reduction in dB SPL to that shown in Figure 3.2.

So, should one use a loudness scale, instead of the inverse square law, to adjust relative virtual auditory distances? A study by Stevens and Guirao (1962) showed estimates of half-loudness to be equivalent to half of the auditory distance. Figure 3.3 shows the resulting dB levels that would occur using a 10 dB scale based on sones, for the same distances shown in Figure 3.2. Compared to the inverse square law example in Figure 3.2, the loudness scale indicates that if the nearest sound source were again perceived to be one foot away at 85 dB SPL, then there would be fewer increments of doubled distance (5 instead of 8) if the noise floor limit were 40 dB SPL, and that 45 dB SPL would correspond to a maximal distance of 45 feet (compared to the maximal distance of 128 feet predicted by the inverse square law at 43 dB SPL).

If judgments of loudness are the primary cues to judgments of distance, then 10 dB would be more appropriate for a general scaling factor in a 3-D sound system. But loudness increments, like intensity, can only operate effectively as distance cues under conditions where other determining cues, such as reverberation, are not present. And practically speaking, this ignores the inherent and relative loudness of each sound source to other sound sources; the calculation of the loudness of each sound source would require a complicated analysis. Recall that determining the loudness of a given *complex* sound is a complicated issue and difficult to predict absolutely; the role of energy within each **critical band** will affect judgments of absolute loudness (recall that a critical band can be simply thought of as one third octave regions spanning the range of hearing). Remember also that the designer of a 3-D sound or any type of audio system cannot predict the intensity at the eardrum of the receiver at the end of the communication chain, since the end user of an audio system will always have final control over the overall sound pressure level.

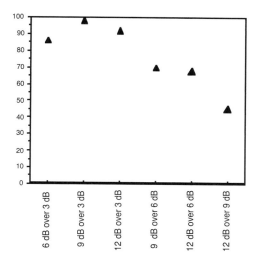

FIGURE 3.4. Results of an informal study (Begault, 1991a) showing preference for producing a sensation of half-distance from a reference. The percentage indicates the proportion of responses favoring the larger dB increase of the paired comparison.

In a pilot study that evaluated preference for half-distance from a reference, Begault (1991a) compared 3, 6, 9, and 12 dB increments. A "two-alternative, forced choice procedure" was used, where each combination of two possibilities (3 versus 6 dB, 3 versus 9 dB, 9 versus 3 dB, 6 versus 3 dB, etc.) was presented. The sounds were speech and a broadband click; the results were averaged for the two types of sounds. Subjects were asked which scheme seemed more convincing for creating a sensation of half-distance; results are shown in Figure 3.4. Further investigation suggested that 3 dB was always too little, and that the choice between 6, 9, or 12 dB was mostly a function of the stimulus type.

However, it has already been pointed out that sounds are not heard in anechoic environments but in conjunction with reverberation and other sound sources. Figures 3.5 and 3.6 show how the influence of reverberation can preclude the use of either the inverse square law or a loudness scale in a 3-D audio simulation. The total sound intensity reaching a listener by direct and indirect paths can be accounted for by using a room simulation program; a simple enclosure is shown in Figure 3.5. The result of calculating the overall dB SPL is shown in Figure 3.6, for anechoic and reverberant conditions. Note how the overall SPL does not change more than about 3 dB between the closest and farthest listening positions, in spite of three doublings of distance! In a reverberant context, the change in the proportion of reflected to direct energy (the **R/D ratio**, explained later in this Chapter) functions as a stronger cue for distance than intensity scaling.

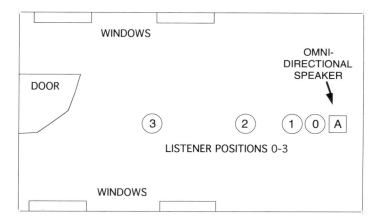

FIGURE 3.5. A modeled "simple room" (dimensions 15 x 8 x 10 feet) used for evaluating sound source intensity in a room. Listener positions 0, 1, 2, and 3 are at 1, 2, 4, and 8 feet, respectively. The modeled room consists of a wooden door, four windows, a carpet on concrete floor, and plywood. *Graphic adapted from Modeler® room design program, Courtesy of Bose® Corporation.*

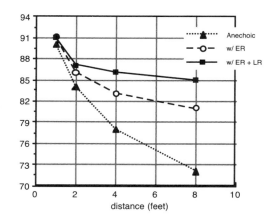

FIGURE 3.6. Reduction in intensity under anechoic and reverberant conditions, at 1, 2, 4, and 8 feet sound source distances, for the room shown in Figure 3.5. ER = early reflections; LR = late reverberation. An omnidirectional sound source is assumed. The solid triangles show the 6-dB reduction in intensity predicted by the inverse square law, under anechoic conditions. The open circles show the reduction in intensity when only early reflections (reverberation up to 80 msec) are taken into account (see reverberation section, below). The filled squares represent the intensity when all reverberant energy is included. Note that the intensity differences under reverberant conditions are minimized (flatter slope of the graph) as distance increases.

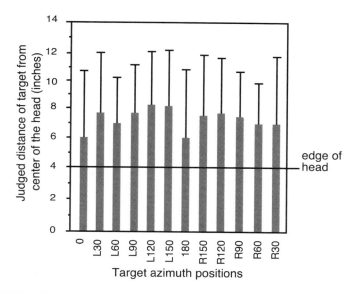

FIGURE 3.7. Means and standard deviations for distance judgments of anechoic speech stimuli, collapsed across the 11 subjects shown in Figure 2.30 (adapted from Begault and Wenzel, 1993). Subjects were instructed to use 4 inches as a reference point from the center to the edge of the head.

What about absolute estimates of distance of HRTF-spatialized stimuli? Figure 3.7 shows a summary of the distance estimates shown previously in Figure 2.30 (Chapter 2) for nonindividualized HRTFs. In that study, speech stimuli recorded under anechoic conditions were played back at a nominal level of about 70 dB SPL; this level corresponds roughly to normal speech at about 1 meter from a listener (Kryter, 1972). Subjects were told to report a distance of the apparent sound image as follows: 0 inches if the sound seemed directly inside the head; greater than 0 but less than 4 inches if the sound seemed displaced inside the head but not at the center; 4 inches exactly for a sound that seemed on the edge of the head; and greater than 4 inches for externalized stimuli.

The results of Figure 3.7 show an overall underestimation of the apparent distance, if the "normal speech" location of 1 meter is to be taken literally. This underestimation has also been observed with actual as opposed to virtual sound sources (Holt and Thurlow, 1969; Mershon and Bowers, 1979; Butler, Levy, and Neff, 1980). One reason for this underestimation may be the absence of reverberation in the stimulus. Underestimation in general may also be related to the bounds of perceptual space, i.e., the **auditory horizon**, discussed later. But note that the standard deviation bars of Figure 3.7 indicate a high amount of variability between subjects, for what was essentially one target distance; one might have expected a more stable distance estimate among individuals, based on the common familiarity of speech. The asymmetry of the azimuth centroids in Figure 2.30, as well as the overall differences between subjects, reveals how distance estimates can vary as a function of the simulated azimuth. This is not to imply that subjects are more accurate in estimating distance from one direction versus another, but it does suggest that different spectral weightings from the various HRTFs used could have influenced judgments. A topic related to absolute distance perception of HRTF-filtered sources, **inside-the-head localization**, is covered later in this chapter.

Influence of Expectation and Familiarity

Distance cues can be modified as a function of expectation or familiarity with the sound source, especially with speech (Coleman, 1962; Gardner, 1969). This is a well-known art to ventriloquists, or to anyone who ordered a "throw your voice" kit from an advertisement in the back of a comic book. "The listener's familiarity with both the source signals and the acoustic environment is clearly a key component in any model for auditory distance perception" (Sheeline, 1982). Hence, any reasonable implementation of distance cues into a 3-D sound system will probably require an assessment of the cognitive associations for a given sound source. If the sound source is completely synthetic (e.g., pulsed white noise), then a listener may need more time to familiarize themselves with the parametric changes in loudness and other cues that occur for different simulated distances. If the sound source is associated with a particular location from repeated listening experiences, the simulation of that distance will be easier than simulation of a distance that is unexpected or unfamiliar. For example, it's easier to simulate a whispering voice six inches away from your ear than it is to simulate the same whisper six yards away.

Gardner (1969) conducted several studies for speech stimuli that illustrate the role of familiarity and expectation. In one experiment, categorical estimations were given by subjects of sound source positions at 0 degrees azimuth, by choosing from numbered locations at 3, 10, 20, and 30 feet in an anechoic chamber. When loudspeaker delivery of recorded, normal speech was presented, the perceived distance was always a function of the sound pressure level at the listener instead of the actual location of the loudspeaker. But with a live person speaking inside the chamber, subjects based their estimates of distance on the manner of speaking rather than on the actual distance. Figure 3.8 shows an illustration of these results. Listeners overestimated the distance of shouting in reference to normal speech and underestimated the distance of whispering, although the opposite should have been true if intensity were the relevant cue.

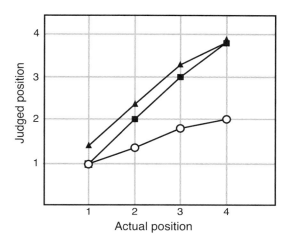

FIGURE 3.8. Results obtained by Gardner (1969) for a live speaker at 0 degrees azimuth in an anechoic chamber. Open circles = whispering; solid squares = low- level and conversational-level speech; triangles = shouting.

Spectral and Binaural Cues to Distance

The spectral content of a sound source relative to a receiver position can vary as a function of its distance. The effects include the influence of atmospheric conditions, molecular absorption of the air, and the curvature of the wavefront. All of these factors modify the spectral content of the sound source in an additive manner, along with the spectral cues of the HRTF. From a psychoacoustic standpoint, these cues are relatively weak, compared to loudness, familiarity, and reverberation cues. One might possibly effect a change in perceived distance by using these cues directly in a 3-D sound system, but experience with several algorithmic implementations suggests that this is not the case. Since the sound sources used in a 3-D sound system are more than likely dynamically changing in location, their spectra are constantly changing as well, making it difficult to establish any type of "spectral reference" for perceived distance. It is worthwhile to bear in mind the following quote from Durlach and Colburn (1978): "As one would expect on the basis of purely physical considerations, if one eliminates the cues arising from changes in loudness and changes in reverberant structure, the binaural system is exceedingly poor at determining the distance of a sound source." Nevertheless, this information is presented for completeness, and may be useful in developing accurate auralization models. For instance, the summed result of air absorption filtering on the multiple reflections of reverberation is probably more important than air absorption effects on the direct sound.

Spectral Changes and Sound Source Distance

A sound source's wavefront is usually planar by the time it reaches a listener's ears, but a closely proximate sound source will have a curved wavefront. This results in an added emphasis to lower versus higher frequencies of the sound source. von Békésy (1960) concluded that "Within the range of roughly a meter, it is apparent that an emphasis of low frequency sound energy, relative to high frequency energy, should indicate the approach of a sound." As a sound source moves beyond a distance of approximately two meters, the wavefront becomes planar rather than spherical, and this rule begins not to apply (Blauert, 1983; von Békésy, 1960; Coleman, 1968). A tangible way of explaining this phenomenon is the "darkening" of tone color that occurs as a sound source is moved very close to one's ear. This effect is related to the equal loudness contours, which show that sensitivity to low frequencies increases with increasing sound pressure level.

This **tone darkening** cue for a nearby source is probably less frequently encountered in everyday listening than the experience of **diminished high-frequency content** that occurs as a sound moves into the distance (Coleman, 1963). With increasing distance, the higher frequencies of a complex sound are increasingly affected by air humidity and temperature. Although Blauert (1983) states that the effect of air absorption on the spectrum of signals above 10 kHz at the ears can become distinctly audible at distances as short as 15 feet, he provides the caveat that "there is still little experimental evidence that the frequency dependence of air attenuation is evaluated in forming the distance of an auditory event."

Results in the literature vary a bit as to what the constants are for calculating absorption coefficients as a function of air temperature and humidity, but information useful for accurate models can be obtained within noise-control references such as those by Harris (1966). An **air absorption coefficient** can be calculated that represents the attenuation of sound as a result of viscosity and heat during a single pressure cycle, and from the interaction of water vapor, nitrogen, and oxygen. Its effect is relatively negligible at short distances, but it becomes more relevant especially for high frequencies at large distances. For example, the data given by Harris (1966) show that at a distance of about 100 meters, an air temperature of 68° Fahrenheit and 20% humidity would cause an attenuation of 7.4 dB at 4 kHz, which is significant; but the attenuation would be less than 1 dB at distances less than 10 meters. Figure 3.9 shows a commonly used set of air absorption curves, and the attenuation that occurs for every 100 feet of distance.

An interesting aspect of air absorption coefficients is that they are in reality a time-varying phenomenon; especially in indoor environments with heating-ventilation-cooling (HVAC) systems, the humidity and temperature at a single point will constantly be in flux, and so will the spectral content of a distant sound source measured at the listener's ear.

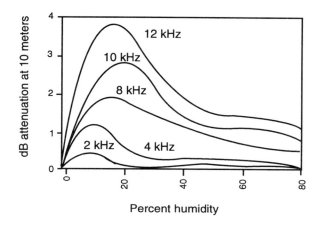

FIGURE 3.9. Absorption of sound in air as a function of relative humidity. Curves are shown for 2, 4, 6, 8, 10 and 12 kHz. After Harris (1966).

From the standpoint of auditory simulation, the spectral cues for distance of a proximate soft sound source and a loud, distant source may be contradictory (Coleman, 1963). The intensity of high-frequency energy is scaled in relation to the intensity of low-frequency energy with both nearby sound sources (closer than 2 meters) and distant sound sources (farther than 15 meters). In the case of distance simulations using headphones, how would such a scaling be interpreted? Butler, Levy, and Neff (1980) processed stimuli with filters of the **low-pass** (i.e., with high frequencies attenuated) and **high-pass** (with low frequencies attenuated) type at several "cut-off" frequencies. They found that the low-pass stimuli were consistently perceived to be farther away than the high-pass stimuli and suggest that "a lifetime of experience" with distant sources helps establish this cue. Perhaps, then, in the absence of other cues, a low-frequency emphasis applied to a stimulus would be interpreted as "more distant" compared to an untreated stimulus.

Frequency Dependent Attenuation at Large Distances

In the environment outdoors, several other factors can contribute to the frequency-dependent attenuation of a sound source with distance along with humidity. These factors can include the effect of wind profiles, ground cover, and barriers such as buildings. The relevant data have been gathered and used primarily by noise-control specialists, who determine what the effects of existing or modified environmental features will have on the dispersion of noise sources, e.g., freeway noise and its containment by "sound walls." This body of information has considerable potential to be included within the modeling of a

Type of attenuation		500 Hz	1 kHz
wind profile:	(downwind)	minimal	minimal
	(upwind)	up to 30 dB	up to 30 dB
ground cover:	short grass	3 dB	3 dB
	thick grass	5.4 dB	7.4 dB
	trees	8 dB	10 dB

TABLE 3.1. Environmental effects on the spectra of a 100 meter distant sound source (after Gill, 1984).

virtual environment. For instance, Gill (1984) has calculated attenuation for a sound source 100 meters distant at 500 Hz and 1 kHz, as a function of wind and ground cover; see Table 3.1. Looking at these data, it seems clear that to create a realistic acoustic mapping of an outdoor virtual environment, the creator of the virtual world would have to know when the lawn was last mown (for suburban virtual reality) and what the current wind direction was at a given moment.

Binaural Cues and Intensity; Auditory Parallax

In an anechoic environment, can one perceive the distance of a sound source more accurately with two ears than one? It is usually assumed that beyond a distance of about one meter, the spectral changes caused by the HRTF are constant as a function of distance. But when one ear is turned at 90 degrees azimuth towards a relatively close sound source, the ITD is greater at lower frequencies than for a sound at a greater distance, as a function of the wavefront no longer being planar but instead spherical as a function of frequency. This has been termed **auditory parallax,** and has been interpreted by some to mean that the accuracy of estimation of a sound from the side should be improved when compared to distance perception on the median plane (Holt and Thurlow, 1969). However, Mershon and Bowers (1979) and Butler, Levy and Neff (1980) found no significant difference in distance localization between frontal and side presentations of a stimulus. In fact, these binaural cues are probably overwhelmed by other distance cues, such as reverberation and loudness, making their inclusion within a 3-D audio system unnecessary. Blauert (1983), citing the numerous discrepancies in the literature, considers the question of binaural cues to distance unresolved.

Inside-the-Head Localization: Headphone Distance Error

As noted in Chapter 2, lateralization studies conducted with headphones result in **inside-the-head localization (IHL)**. Experiment for a moment by talking with your hands over your ears, and then off; the sound image will noticeably jump to an inside-the-head position. IHL also occurs with 3-D sound techniques, especially without reverberation. This is particularly important, since headphone playback is otherwise superior to loudspeakers for transmitting virtual acoustic imagery in three dimensions; usually loudspeaker sound sources are heard outside the head. But sometimes actual external sources yield IHL, and some sounds can be placed outside and then inside of the head at will, as with the Necker cube illusion shown previously in Figure 2.4. Hanson and Kock (1957) obtained IHL with two loudspeakers in an anechoic chamber, each playing the same signal but 180 degrees out of phase.

Durlach and Colburn (1978) have mentioned that the externalization of a sound source is difficult to predict with precision, but "clearly, however, it increases as the stimulation approximates more closely a stimulation that is natural." The likely sources of these natural interaural attributes include the binaural HRTF, head movement and reverberation. Many researchers have discounted theories that IHL is a natural consequence of headphone listening (due to bone conduction or pressure on the head), simply because externalized sounds *are* heard through headphones in many instances.

The goal of eliminating IHL effects arose in the 1970s with the desire to make improved binaural (dummy head) recordings. Many who heard these recordings were disturbed by the fact that the sound remained inside the head, as with lateralization. Ensuring that the sound was filtered by accurate replicas of the human pinnae and head was found to be an important consideration. Plenge (1974) had subjects compare recordings made with a single microphone to those made with a dummy head with artificial pinnae; the IHL that occurred with the single microphone disappeared with the dummy head micing arrangement. Interestingly, this study also showed that IHL and externalization can occur simultaneously. Laws (1973) and others determined that part of the reason for this had to do with non-linear distortions caused by various parts of the communication chain, and the use of **free-field** instead of **diffuse-field** equalized headphones (see Chapter 4 for more on diffuse-field equalization).

In virtual acoustic simulations, IHL may be *more likely* to occur when head movement is not accounted for, probably for the same reason that reversals occur (Durlach, et al., 1992). The ability of head movement cannot be used to disambiguate locations (see discussion of the cone of confusion and reversals in Chapter 2), which can potentially lead to judgments at a "default" position inside or at the edge of the head (see also Wallach, 1940). There has been no systematic study on the contribution of head movement to externalization, as of this writing. If visual cues are supplied—e.g., if one can walk away or towards a

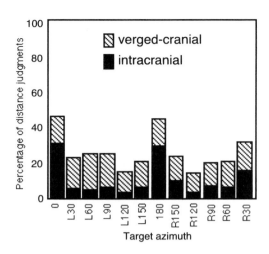

FIGURE 3.10. IHL speech stimuli as a function of azimuth; results from 11 subjects (Begault and Wenzel, 1993). *Reproduced with permission from Human Factors, Vol. 35, No. 2, 1993. Copyright 1993 by the Human Factors and Ergonomics Society. All rights reserved.*

virtual acoustic source in a fully immersive virtual environment—then it is quite likely that the combination of vestibular and visual cues will enable externalization. But note that, due to the importance of visual capture, externalization is also possible listening to a television with a single earpiece.

Figure 3.10 shows the results obtained by Begault and Wenzel (1993) for 3-D processed speech stimuli under anechoic conditions. This figure shows a closer look at the distance judgments shown in Figure 2.30. "Verged-cranial" and "intracranial" refer to judgments at the edge of and inside the head, respectively. The mean percentage of unexternalized stimuli was about 33%, with the worst cases being for 0 and 180 degrees azimuth, positions with the least interaural differences. However, in a study by Wenzel *et al.* (1993) that used the same HRTFs (and interpolated versions), it was found that white noise bursts are almost always externalized. This result suggests on one hand that externalization may depend on the sound source itself and the manner in which it interacts with the HRTF; white noise contains a broad frequency spectrum affected by the entire frequency range of the HRTF, while speech is relatively band-limited at high frequencies. But on the other hand, familiar sound sources (e.g., speech) have been cited as yielding more accurate distance estimates than unfamiliar sources such as noise (Coleman, 1962; McGregor, Horn and Todd, 1985). Perhaps, with the speech experiment, the sources were underestimated due to the unfamiliarity of the anechoic conditions (Gardner, 1969).

Subject Number	IHL Anechoic	IHL Reverberant
1	46%	10.5% ***
2	59.5%	2% ***
3	3.5%	0.5%
4	-0-	0.5%
5	15.5%	-0- ***
Mean	**25%**	**2.7%**

TABLE 3.2. Data summarizing IHL judgments of virtual speech, for five subjects under dry and reverberant conditions (adapted from Begault, 1992a). Columns 2 and 3 show percentages of all judgments by each subject that were unexternalized—heard less than or equal to 4 inches distance. The stars indicate a very high (> 99%) probability that the difference between anechoic and reverberant listening was significant.

Researchers have almost consistently found that reverberation, either natural or artificial, enhances the externalization of 3-D headphone sound (Plenge, 1974; Toole, 1970; Sakamoto, Gotoh, and Kimura; 1976; Begault, 1992a). Thus, it would seem that in a virtual acoustic environment, the inclusion of not only direct paths but indirect paths of reflected sound is necessary for a realistic simulation. Table 3.2 shows distance data from an experiment where subjects compared anechoic and reverberated 3-D processed speech stimuli (Begault, 1992a). The simulated reverberation was also HRTF filtered, using auralization techniques covered in Chapter 4. Columns 2 and 3 show IHL judgments as a percentage of the total number of judgments for each subject. Overall, IHL was reduced from 25% to almost 3%, a substantial improvement. Immediately apparent is the high degree of variability among the five subjects; subjects 1 and 2 heard close to half of the stimuli with IHL, while subjects 3–5 have much smaller percentages (0, 3.5 and 15.5%). Adding synthetic reverberation thus helped the "IHL prone" subjects.

REVERBERATION

Physical Aspects of Reverberation

In most descriptive models of reverberation, a sound source's **direct sound** is defined as the wavefront that reaches the ears first by a linear path,

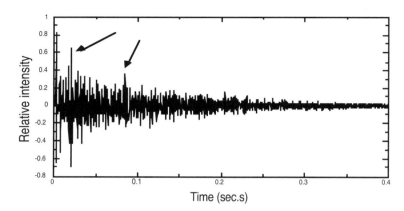

Time (sec.s)

FIGURE 3.11. An impulse response measured in a classroom with an omnidirectional microphone. Arrows indicate significant early reflections.

without having bounced off a surrounding surface. **Reflected sound,** i.e., **reverberation**, refers to the energy of a sound source that reaches the listener indirectly, by reflecting from surfaces within the surrounding space occupied by the sound source and the listener. This surrounding space is a type of **environmental context**, as described in Chapter 1 (see Figure 1.2). Only in **anechoic chambers** or in atypical environmental conditions, such as within a large, open expanse of snow-covered ground or on a mountain summit, are sound sources ever non-reverberant. If there are no obstructions between the sound source and the receiver, the direct sound can be considered to be equivalent to a sound within an anechoic chamber.

Figure 3.11 shows an **impulse response** of a real room, obtained by recording a loud impulsive noise, such as a starter pistol being fired, with an omni-directional measurement microphone. An impulse response is a useful time domain representation of an acoustical system, including rooms and HRTFs. A dummy head binaural micing system, or probe microphones located in the ears of a listener, would reveal the convolution of a particular set of pinnae with the impulse, resulting in a **binaural impulse response.** Chapter 4 covers concepts and modeling techniques of room, HRTF, and binaural impulse responses in greater detail.

Figure 3.11 also shows identification of two possibly significant **early reflections**. A particular reflection within a reverberant field is usually categorized as an **early reflection** or as **late reverberation** (i.e., late reflections), depending on time of arrival to the listener. Significant early reflections, those with a significant amplitude above the noise floor, reach the receiver within a period around 1–80 msec, depending on the proximity of

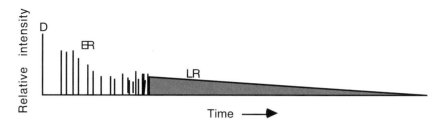

FIGURE 3.12. Simplified reflectogram of an impulse response. D = direct sound; ER = early reflections, > 0 and < 80 msec; LR = late reflections, > 80 msec). The late reverberation time is usually measured to the point when it drops 60 dB below the direct sound.

reflecting surfaces to the measurement point. The early reflections of a direct sound are followed by a more spatially diffuse reverberation termed **late reverberation** or **late reflections**. These later delays result from many subsequent reflections from surface to surface of the environmental context. This temporal portion of the impulse response contains less energy overall.

Reflected energy is categorized in terms of early and late reflections due to both physical and psychoacoustic criteria. In a heuristic "theoretical" room, the buildup of successive orders of reflections begins to resemble an exponentially decaying noise function during the late reverberation period, causing individual reflections to be lost in the overall energy field. Psychoacoustically, the individual earlier reflections are less likely to be **masked** (inaudible) than later reflections. The psychoacoustic evaluation of reverberation places most of its emphasis on the first 80 msec of the response, although this cut-off point is largely arbitrary.

In reverberation models, early reflections are usually modeled directly, by tracing sound wave paths in a manner akin to tracing light. This is a valid procedure for plane waves that are smaller than the incident surface. If a **ray tracing** or **image modeling** algorithm (described in Chapter 4) is used to calculate reflection paths from source to listener, a more abstract representation of an impulse response can be used as shown in Figure 3.12. This type of temporal picture is sometimes referred to as a **reflectogram**. Due to the increase in density and corresponding reduction in intensity of late reflections over time and the finite limits of computation power, it is common not to model each individual reflection in the late reverberation period, but instead to model them using a Gaussian distribution with decaying envelope.

Figure 3.13 shows how the calculation of the time of arrival and angle of incidence of a single reflected sound is taken into account. As will be seen in Chapter 4, the modeling of an entire impulse response is an extension of this technique. The angle of incidence to the wall equals the angle of reflection to

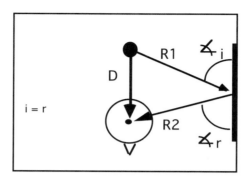

FIGURE 3.13. Sound reflection off of a reflective surface in an anechoic chamber. D = direct sound (angle of incidence to listener is 180°). R1 and R2 are the reflected sound paths. The reflection angle of incidence equals the angle of reflection. The reflection arrives at the listener at an angle of about 120° left.

the listener; and the angle of incidence from the wall to the listener relative to the listener's orientation determines the spatial component of the reflection. Multiple reflections will arrive from many directions, but in many circumstances the significant early reflections come from a limited set of directions. The temporal aspect of the reflection is derived from the distance it must travel, divided by the speed of sound in the enclosure. Finally, an overall attenuation of the reflection will occur as a function of the inverse square law.

Each successive bounce of a reflection off a surface is referred to as an *increase in reflection order*. Hence, first-order reflections have bounced off of one surface before reaching the receiver; second-order reflections have bounced from one surface and then to another surface before reaching the receiver; and so on. The intensity of the reflection is progressively reduced by each successive bounce since the material of the wall will absorb a portion of the total energy in a frequency-dependent manner. However, the number of low-amplitude reflections is quite large, and gets larger with each successive order, causing the combined effect of late reverberation.

Figure 3.14 shows a graph from a room design program that shows the increasing density of first-, second- and third-order early reflections. Figure 3.15 contrasts the pattern of early and late reflected energy in a small room and a large room. Note that in the smaller room, the energy builds and decays rather quickly, while in the large room the buildup and decay are much slower. Note also that more of the early reflections are masked in the buildup of the larger room than in the smaller room.

Early and late reflections are often described collectively, by the following three physical parameters: its duration, usually evaluated as **reverberation time**, or **t60**; the ratio of reverberant-to-direct sound, the **R/D ratio**; and in terms of several criteria related to the arrival time and spatial incidence of early reflections.

The first parameter, reverberation time (t60), is defined as the duration necessary for the energy of reflected sound to drop 60 dB below the level of the direct sound. The value of 60 dB is somewhat arbitrary, since reverberation can be significant to the level that it is still audible; its original meaning was to

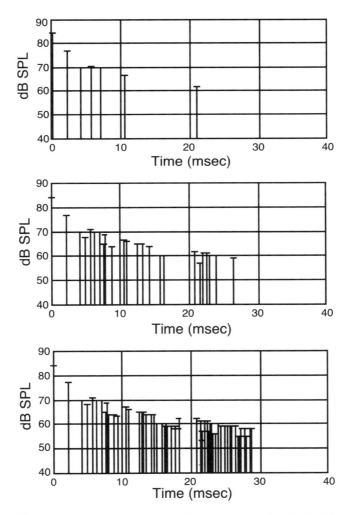

FIGURE 3.14. Reflectograms from a modeled room, showing the buildup of early reflection density with increasing reflection orders. Top: first order; middle: first and second order; bottom: first, second and third order. *Graphic adapted from Modeler® room design program, Courtesy of Bose® Corporation.*

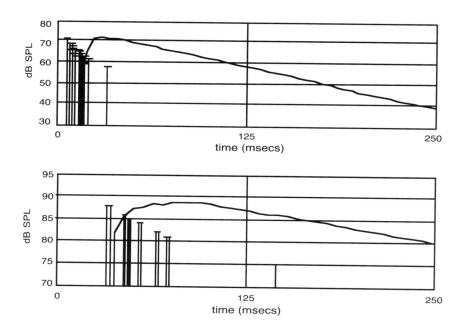

FIGURE 3.15. Reverberant decay and early reflections in (top) a small room 2 feet from a sound source, and (bottom) in a very large room, 30 feet from a sound source. Note that in the larger room the early reflections are masked by the reverberant decay and that the rate of decay is much slower than in the smaller room (7 dB between 125—250 msec, as opposed to 19 dB). *Graphic adapted from Modeler® room design program, Courtesy of Bose® Corporation.*

simply indicate a value for "relative silence." If the loudest sound in an environment was 90 dB and the ambient noise was 40 dB—not atypical for many environments—then t60, as opposed to t90, would be adequate measure, because the reverberation below 40 dB would be masked by the noise floor. In terms of the listening experience, t60 is related to the perception of the size of the environmental context; a larger value for t60 is usually associated with a larger enclosure.

The second parameter, the R/D ratio, is a measurement of the proportion of reflected-to-direct sound energy at a particular receiver location. As one moves away from a sound source in an enclosure, the level of the direct sound will decrease, while the reverberation level will remain constant. This is interpreted primarily as a cue to distance (see discussion ahead). At a point where the reflected energy is equivalent to the direct energy in amplitude, the term **critical distance** is used.

The third parameter is related to temporal and spatial patterns of early reflections. The spatial distribution of early reflections over time in a given situation exist as a function of the configuration between sound source, listener, and physical features of the environmental context. Different spatial-temporal patterns can affect distance perception and image broadening, as well as perceptual criteria that are related to the study of concert halls (Begault, 1987; Kendall, Martens, and Wilde, 1990). By measuring the similarity of the reverberation over a specific time window at the two ears, a value for **interaural cross-correlation** can be obtained (Blauert and Cobben, 1978; Ando, 1985).

Perceptual Aspects of Reverberation

Relevant physical parameters of reverberation that affect the perception of an environmental context include the volume or size of the environmental context, the absorptiveness of the reflective surfaces, and the complexity of the shape of the enclosure. The size of the environmental context is cued by the reverberation time and level. The absorptiveness of the reflective surfaces will be frequency dependent, and will vary from environmental context to environmental context, allowing for cognitive categorization and comparison on the basis of timbral modification and possibly on the basis of speech intelligibility. Finally, the complexity of the shape of the enclosure will shape the spatial distribution of reflections to the listener, particularly the early reflections. It is this latter area for which a considerable literature exists in the domain of concert hall acoustics (see, e.g., Beranek, 1992).

Late Reverberation

Late reverberation is informative perceptually as to the volume of a particular space occupied by a sound source. This is most noticeable when a sound source stops vibrating (i.e., is "turned off"), since one can hear the time it takes for the later echoes to decay into relative silence, and the shape of the amplitude decay envelope over time. The long-term frequency response of a room—i.e., the reverberation time for each one third octave band—can also be significant in the formulation of a percept of the environmental context. During continuous speech or music, only the first 10–20 dB of decay will be heard, since energy is constantly being "injected" back into the system.

The first concert hall studies at the beginning of the century emphasized reverberation time as a strong physical factor affecting subjective preference. In essence, however, t60 gives a very limited description of the reverberation amplitude envelope over time. Griesinger (1993) and others have pointed out the usefulness of measuring reverberant **level** over time, as opposed to just t60. It is even better to analyze the late reverberation level as a function of time within each critical band, or its approximation as one third octave bandwidths across the entire audible range. The distinguishing characteristic between many concert halls and other reverberant environments lies in the relative decay pattern of each of these frequency regions. Many of these regions do not decay exponentially, although this is a standard approximation used to describe what happens "between" time 0 and t60.

The R/D ratio and Distance Perception

The reverberant-to-direct sound (R/D) ratio has been cited in many studies as a cue to distance, with varying degrees of significance attributed to it (Rabinovich, 1936; Coleman, 1963; von Békésy, 1960; Sheeline, 1982; Mershon and King, 1975; Mershon and Bowers, 1979). Sheeline (1982) found that reverberation was an important adjunct to intensity in the formation of distance percepts, concluding that "reverberation provides the 'spatiality' that allows listeners to move from the domain of loudness inferences to the domain of distance inferences."

Von Békésy (1960) observed that when he changed the R/D ratio, the loudness of the sound remained constant, but a sensation of changing distance occurred. He noted that "though this alteration in the ratio between direct and reverberant sound can indeed be used to produce the perception of a moving sound image, this ratio is not the basis of auditory distance. This is true because in [an anechoic room] the sensation of distance is present, and, in fact, is even more distinct and of much greater extensiveness than elsewhere." He also observed that the sound image's width increased with increasing reverberation: "Along with this increase in distance there was an apparent increase in the size or vibrating surface of the sound source... for the direct sound field the source seemed to be extremely small and have great density, whereas the reverberant field had a more diffuse character." While direct and indirect sounds reach a listener in different proportions as a function of their distance, it is true that in some contexts, the possible variation in the R/D ratio can be limited by the size of the particular environmental context, causing the cue to be less robust. For instance, in a small, acoustically treated room, the ratio would vary between smaller limits than in a large room like a gymnasium.

Figure 3.16 and Table 3.3 show the overall effect of adding a particular configuration of synthetic reverberation to 3-D audio speech, as a function of virtual azimuths (mean values for five subjects). These data are from the same study as shown in Table 3.2. The modeled virtual source–listener configuration had a distance of about 39 inches; reverberation can be seen to improve the simulation and allow externalization. While Table 3.3 shows that the overall effect of reverberation varies widely between individuals in terms of the

Subject	Anechoic	Reverberant	Increase
1	5	13.1	2.6:1
2	5	14.5	2.9:1
3	14.9	34.9	2.3:1
4	26.8	90.4	3.3:1
5	15.3	59.4	3.8:1
mean	13.4	42.5	3.2:1

TABLE 3.3. Data for distance judgments for five subjects under dry and reverberant conditions. Columns 2 and 3 show mean distance estimates in inches, along with the relevant increase in distance shown as the ratio of reverberant to dry estimates (Begault, 1992a).

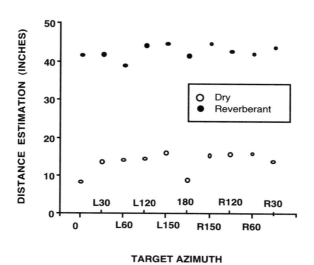

FIGURE 3.16. Mean distance estimates from five subjects (see Table 3.2) for dry and reverberant stimuli (Begault, 1992a). While the variation between individuals for absolute distance estimates was high, the proportion of the increase from their dry estimates ranged between 2.3–3.8 to 1.

absolute number of inches reported, it also shows that the relative judgment of distance from the initial estimate of the anechoic stimuli ranged only from 2.3–3.8 to 1 between individuals, for the particular reverberant stimulus used (Begault, 1992a).

The Auditory Horizon

Sheeline (1982), like von Békésy (1960), found that reverberation defines a boundary for distance judgments; there is an upper limit to how much reverberation can be mixed with a sound before it reaches an "auditory horizon." Sheeline's study also shows that distance perception is improved under optimal reverberation conditions relative to anechoic playback. This is contrary to von Békésy's observation that distance perception is "more distinct" and "of much greater extensiveness" in anechoic conditions than within reverberant contexts.

The reverberation of an environmental context can serve as an aid in identifying a likely range of sound source distances. Mershon and King (1975) comment on the contribution of reverberant energy as allowing the listener to set the "appropriate boundaries (or potential boundaries) for a source or a set of sources." If reverberation serves as a cue to the extent of the environmental context's size or characteristics, particularly the distance to the boundaries of surfaces, then the cognitive process, as informed by audition and any other available senses, could cause perceptual limits to the possible extent of sound source distances.

An everyday example of the effect is the sense of "intimacy" forced upon people in small enclosures such as an elevator or a jet airliner. Someone talking to you in this type of enclosed space seems closer than when in a large space (for instance, an exhibition hall), although the actual physical distance is the same. A preliminary investigation by Begault (1987) lends some support to this concept. With loudness-equalized speech sources processed only with spatialized early reflections, subjects generally found that a virtual sound source at two meters distance seemed closer in a "large" modeled enclosure than a "small" modeled enclosure. The physical difference between these two conditions was that the early reflections were spread out over a greater time period and were weaker relative to the direct sound than in the small enclosure. The spectral change resulting from the two different reflection patterns was assumed to be the primary cue responsible for the effect.

Effect of Reverberation on Azimuth and Elevation Estimates

While the presence of reverberation can improve externalization of virtual acoustic images, it can also decrease localization accuracy under real and simulated conditions. This is likely to happen in situations where the strength and timing of early reflections is such that the attack portion of the amplitude

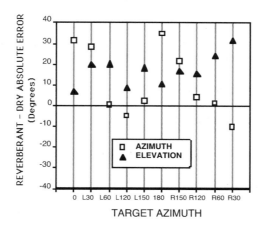

FIGURE 3.17. Difference in absolute error for azimuth and elevation between reverberant and anechoic stimuli (Begault, 1992a).

envelope of a sound is "smeared" in time and space. In fact, a desirable effect of the reverberation in concert halls is to blend individual instruments that would otherwise stand out.

In a localization study using the variable reverberation acoustics of an adjustable room (Hartmann, 1983), it was found that lateral early reflections (reflections from the side) degraded localization accuracy. The expanded apparent source width that can occur with reverberation could also make the estimation of a particular location more difficult, since there is a larger area on which to consistently place a center of gravity for azimuth (and elevation) judgments. Figure 3.17 shows the difference in localization accuracy for HRTF-processed speech between anechoic stimuli and stimuli processed with synthetic spatialized reverberation (Begault, 1992a). For most target azimuths, localization is worse with the addition of reverberation. This demonstrates how the precedence effect only partially suppresses the effects of reflected energy.

Specific Perceptual Effects of Early Reflections

Echolocation

Echolocation, the effect of an initial reflection on perceived pitch or timbre, has been discussed as a feature the blind can use to determine perceived distance from a surface. At one time it was suspected that the blind may have had a sixth sense of "facial vision," related to a tactile sensation felt on the face before encountering an object (Cotzin and Dallenbach, 1950). It was later found that both blind or sighted blindfolded subjects could make use of clicking or hissing sounds from the mouth to estimate the distance, width, and even material composition of objects placed in the ± 90-degree frontal arc (Kellogg, 1962; Bassett, 1964; Rice, Feinstein, and Schusterman, 1965; Rice, 1967). This is accomplished by noticing subtle timbral changes that occur with small time delays, or to other acoustic phenomena as explained in connection with the precedence effect (see Chapter 2; Figure 2.5).

Some persons are better at echolocation than others, and most people tend to improve when they can choose their own sonar signal. Distance differences as small as 4 inches can be perceived from the reflected sound from a disc one foot in diameter placed about 2 feet away; this corresponds to a time-delay range of about 4–6 msec. One amazing result found by Kellogg (1962) with two blind subjects was their ability to discriminate between discs covered with hard and soft surfaces (e.g., glass versus wood); subjects could discriminate between discs covered with denim and with velvet 86.5% of the time! These results indicate the level to which hearing acuity can be developed; there is no reason to believe that such an ability could not eventually be made relevant to an everyday user of a 3-D sound system connected to, for instance, the user's office computer.

Timing and Intensity

The timing and intensity of early reflections can potentially inform a listener about the dimensional aspects of the environment immediately surrounding the sound source. Picture listening to sounds late in the night from your house or apartment before falling asleep. You'll notice that when it's impossible to identify specific sound sources, their direction or distance, the reverberation gives clues for categorizing their location according to previous knowledge of the environment.

Clearly, the formation of a particular spatial percept is a result of not only the direct sound, but its fusion with early reflections, as exemplified by the effects of precedence, masking, and temporal integration. Early reflections are not usually heard as distinct, separable sound sources, but instead are *indirectly* perceived, in terms of their overall effect on a sound source. A notable exception are reflections that are delayed long enough to be heard as distinct echoes. In everyday listening, the relative intensity and temporal pattern of early reflections can influence the timbre of speech, and even its intelligibility level. The presence of echoes can cause a disastrous situation for public address or telecommunication systems; the design of physical and electronic compensation methods is a frequent engineering activity in the design of conference rooms, churches, and spaces with large sound system installations. This has resulted in a number of room modeling approaches that are implemented within computer aided-design (CAD) programs, described in detail in Chapter 4.

The temporal resolution for integration of speech or music is around 20 msec, which "irons out" most of the spectral irregularities of a room caused by reflections (Schroeder, 1973). Nevertheless, the perceived timbre and attack qualities of sound sources, especially of musical instrument sounds, can be changed substantially by early reflections without affecting intelligibility. One hears a cello, for example, not only from vibrations emitted from the instrument itself, but also from early reflections from the floor and walls. Most professional musicians are very aware of the manner in which a stage or other floor surface interacts with their instruments. Besides altering the timbre of the instrument, early reflections can also modify the rise time of the attack portion of the amplitude envelope. An extreme example of this is the lack of "bite" a pizzicato (finger pluck) note has on a stringed instrument in a highly reverberant room. The influence of early reflections has also been explored in connection with concert hall acoustics, as a means of analyzing what physical parameters are responsible for making "good-sounding" halls (see below). In this regard, the spatial incidence of early reflections to the listener becomes important.

Concert Hall Reverberation

Surprisingly, consideration of the environmental context by using reflected sound as a variable parameter has only been recently explored in psychoacoustic studies of distance and azimuth perception (Mershon and King, 1975; Sheeline, 1982; Hartmann 1983; Begault, 1992a). Concert hall reverberation studies are where most of the work has been applied to the perceptual effects of reverberation, particularly of early reflections. These studies can be viewed as investigations of a very specific type of environmental context, characterized by a condition where the distance from source to listener is rather long. In addition, only a few seating positions (e.g., "the best seat in the house") are evaluated. The character of the reverberation is also restricted in these studies, since it must favor music of the western "art music" tradition from about 1820 to present. For example, "classical music" written before the nineteenth century was heard within a completely different environmental context, such as a parlor or a church.

As a result, the criteria used for evaluating reverberation and the timing and amplitude of early reflections according to changes in the modeling of the enclosure are toward a rather specific goal. Frequently, concert hall reverberation studies seek to establish methods for how to build a new hall (an infrequent event) or modify an existing hall so as to have ideal characteristics for music, with ideal reverberation time and patterns of reflections. Certainly, the idealized state for the production of virtual worlds would include concert halls — but as only one part of the repertoire of available environmental contexts.

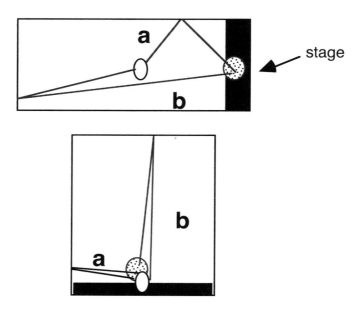

FIGURE 3.18. Simplified plan and section of the "shoe box" concert hall shape. The listener is represented by the oval, the sound source by the dotted circle. "Good" reflections from the side (a) arrive before "bad" reflections from overhead or on the median plane (b).

The temporal and spatial aspects of early reflections have been found to be especially significant in forming preference criteria between various concert halls. The overall finding is that lateral reflections (those near ± 90°) should dominate over reflections that come from the front or the rear (those near 0 and 180°). This dominance can be established through arrival time, direction, and level. Put another way, people seem to be able to associate a "good hall" with one that seems more "stereo" than another (Schroeder, 1980). Lateral reflections, i.e., those that contain slight left–right time differences for a "stereo" effect, are likely to arrive before reflections on the median plane of a listener when the distance to the side walls is shorter than the distance to the front or rear walls, or to the ceiling. The type of performance venue that accomplishes this most well is the classic shoe box design (see Figure 3.18), exemplified by Boston Symphony Hall. The success of the design is proven by the fact that most successful concert halls continue to be built on this model. On the other hand, multipurpose, wider-than-longer modern halls sound worse than a

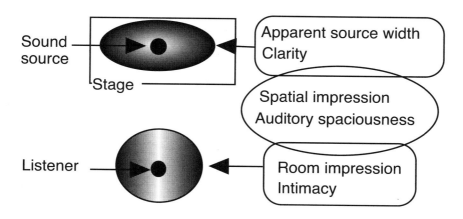

FIGURE 3.19. Perceived auditory spaciousness and apparent source width. An example of apparent source width would be the perceived image size of single cello notes experienced by a nearby listener (or microphone pair) in a reverberant concert hall. Higher-frequency pitches, such as those produced by a bowed open harmonic on the highest string played fortissimo, will appear more spatially defined (the black dot) compared to the lowest note of the cello (an open C) played pianissimo. The same two notes will change their relative extent, i.e., their apparent source width, as a function of the early reflections within the environmental context. However, a change in the extent of all pitches, such that the listener felt more "enveloped" would be described as a change in "intimacy," or "room impression."

shoe box design because the lateral stereo-producing reflections arrive later in time relative to the monaural reflections from the ceiling and front and back walls.

The **apparent source width** of a sound source is defined as its perceived extent or size of the sound source's image, and is related to the concept of **auditory spaciousness**. Blauert (1983) gives the following description of auditory spaciousness: "...perceived spatial extension of the sound source; extension of the region that the sound seems to come from; and spatial extent of the auditory event." In other words, it can be thought of as the opposite of spatial definition, where "maximum definition" is defined as the minimum possible image localization blur of the sound source within all spatial dimensions. Figure 3.19 shows levels of apparent source width graphically for only two dimensions; spaciousness could be extended to elevation extent as well.

Due to the common dependence on early reflection temporal-spatial patterns, the percepts of spaciousness and apparent source width are often mixed with physical measurements of **spatial impression,** which is a physical measure of the relative interaural similarity of a room during the early reflection period from 0–80 msec (Barron, 1971; Barron and Marshall, 1981; Ando, 1985). The perception of spatial impression is apparently equivalent to that given by Blauert (1983) for auditory spaciousness with an additional distance effect; spatial impression provides "a certain degree of envelopment of the sound and [gives] an impression of distance from the source." This has to do with a general impression of a nondirectional "wash" of reverberation about a listener, independent of the apparent source width or direction. Some authors now refer to this as **room impression** (Griesinger, 1993), while Beranek (1992) calls it **intimacy**.

The difficulty in establishing distinctive perceptual criteria in a concert halls study can be quite difficult. In many cases, the match between physical and psychophysical parameters are blurred. In fact, it wasn't too long ago that people rated concert halls with a number of adjectives more appropriate to an informal wine tasting evaluation than to a scientific study (for instance, one study used subjective ratings between adjectives such as "overbearing" and "reticent"!—see Schroeder, 1984). When physical parameters such as early reflections are varied, the criteria used for perceptual judgment have to be understood by all listeners in the experiment, and the individual's own sense of that criteria must remain stable during evaluation.

Some have tried to simplify the situation by limiting the number of perceptual criteria. Borish (1986) stated that "the only subjective effect of delayed sounds when laterally displaced is spatial impression," and that "there is an infinite number of ways of generating the same spatial impression by simultaneously varying the azimuth and amplitude of the secondary sounds. Two sound fields may differ objectively, yet have the same subjective effect." Another example is the concept of **clarity** or **c80**. This is quantified by dividing the energy of the first 80 msec (the early reflections) by the reverb after this point (the late reflections). Griesinger (1993) points out that its perceptual meaning can be ambiguous, implying intelligibility to some people and apparent source width as it relates to the "ease of localization" of a sound source to others. It is evident that a number of inter-related variables are at work that are difficult to pin down. For instance, apparent source width can be affected by spatial impression and clarity, which are in turn related to distance and envelopment by a sound source, all of which are influenced by reverberation.

Other experiments have tried to extract perceptual criteria in a more indirect, exploratory manner. In such experiments, subjective responses are gathered either using preference evaluations or via similarity judgments, which are in turn analyzed by the statistical techniques of correlation analysis and multidimensional scaling (MDS) analysis. In a preference experiment, subjects merely choose which hall in a paired comparison "is preferable," using any criteria they like. Post-analysis involves correlation of identifiable physical features with preferred and non-preferred halls (see Thurstone, 1927; Schroeder,

1984; Ando, 1985). A more complex means of analyzing the relationship between objective criteria and perception is underway at the Institute for Research and Coordination of Acoustics and Music (IRCAM). In these studies subjects rate the apparent similarity of a set of paired simulated and actual halls, each of which differ in some type of objective criteria. A set of 11 separable perceptual factors result from the MDS analysis, for which varying sensitivities can be assigned (Warusfel, Kahle and Jullien, 1993).

If one examines a standard off-the-shelf reverberators, one can find a "concert hall" setting on just about all of them. They will all sound different, except for having a significant amount of reverberation time—i.e., the "same subjective effect." Clearly, considerable work remains to be done in this promising area, before "perception knobs" can be attached to an auralization 3-D sound system; allowing a user to adjust, for example, the sound source in the virtual concert hall to have 10% more apparent source width but 15% less spaciousness.

CHAPTER FOUR

Implementing 3-D Sound

Systems, Sources, and Signal Processing

CONSIDERATIONS FOR SYSTEM IMPLEMENTATION

A successful implementation of 3-D sound is a bit of a juggling act. If resources such as time or money were unlimited and the final product were to be used by a single person, then the task of implementation would be much easier. But usually an aggregate assessment of various needs must be conducted in the early phases of design in order to develop a reasonable system that is both within budget and applicable to a group of people rather than an idealized individual. There will also be cases where sound is included or enhanced within an existing product, presenting an additional set of requirements that need to be balanced. The "audio guru" on a design team for an application in most cases will face the following three issues head-on.

1. From the human perspective, what is the proposed **use of sound** within the application; how will sound motivate, guide, alert, immerse, etc. the user? Are the sounds representational of visual events and therefore tied to visual media? In virtual reality and multimedia applications, most sounds are somehow connected to visual events, requiring decisions as to how representative they need to be. A related question is whether or not a particular sound will act as a carrier for more complex meanings beyond simple representation. Using a desktop flight simulator as an example, the sound of the engine can be modeled to suggest the complex nuances of the engine's current status as a function of the user's operation and possible failures; or a generic "engine sound" can be used that has nothing to do

with the user's operation of the controls. The former is necessary if the end application is intended for professional pilots, and the latter is acceptable for a children's game. How are the sounds generated? How detailed is the virtual audio world that is envisioned—how many sound sources, how many can occur at once? Also, do spatialized sounds need to be produced in **real-time**, or can the sounds be **pre-spatialized**? The answers to these questions are best flushed out into several malleable versions to adhere to the adage "one should have a plan so that there's something to change." Design solutions can be tested against perceptual requirements and then be recycled and morphed (or possibly transmogrified!) via iteration into to more practical versions.

2. The use of sound within the application translates directly into the **perceptual simulation requirements** of the system. It's necessary at this point to make a reality check on the viability of the perceptual requirements, based on what is known on the topic, as described in Chapters 2 and 3. For instance, it would be very difficult to enable a person to localize elevation of a virtual source any more accurately than he or she is able to under normal spatial hearing conditions; and it's known that the HRTFs of the user, if available, will yield better results than nonindividualized HRTFs. The latter is related to the degree that a system can be **perceptually calibrated** to a particular user. Perceptual simulation requirements also will guide the design and the degree of complexity necessary for creating the **sound sources** that are to be spatialized.

3. Finally, perceptual requirements must be balanced against **available resources**. This process not only involves realistic limits imposed by time and monetary budgets, but also the technological aspects of the audio system as a whole. A designer must envision the nature of the **sound system integration requirements**, in terms of hardware and software. This task is partly an inventory of computer resources (memory, available instructions) and of external hardware, such as the effectors and sensors already in use for the visual displays. For example, if a dedicated DSP chip already resides on a computer, one may want to use an **integrated approach** by harnessing some portion of its signal-processing capabilities. There also may be one or more off-the-shelf sound cards or specialized signal processing cards that plug into the bus slots of the computer. On the other hand, the computer may be used simply to trigger sound events to external hardware devices, including spatialization hardware. This use of external hardware devices is referred to as a **distributed approach**.

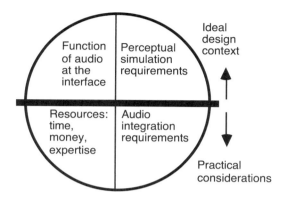

FIGURE 4.1. A requirements versus resource allocation pie chart for designing audio subsystems within virtual reality or multimedia. An argument for the inclusion of 3-D sound processing typically must emphasize perceptual simulation requirements.

Figure 4.1 summarizes these design considerations in pie chart form. At the top are the two areas that, within an ideal design context, would occupy very large portions of the design effort's "planning pie." At the bottom are the practical considerations that are certain to influence the outcome of any 3-D sound implementation within a larger system. Ideally, one would prefer to concentrate all of the effort on the function and perceptual requirements; sometimes, especially if human safety is at stake, these areas can and must be prioritized regardless of practical considerations. The challenge is to adapt to each of these areas into the design process, determine their priority, and predict the future with enough skill to know how big each pie slice ought to be. Finally, as with all design processes, a periodic reevaluation of the planning approach is necessary.

AUDIO SYSTEM TAXONOMY

Hardware and Software Integration Requirements

As pointed out in Chapter 1 and in Figure 1.7, the role of computing devices within a virtual reality or multimedia system can be segregated between the **audio subsystem** and the virtual reality or multimedia **operating system.** The latter includes the host computer's central processor, which contains an active image of the scenario generated by the reality engine software. The scenario can involve software calls on the level of anything from making the operating system's built-in beep respond to an input (e.g., the arrival of

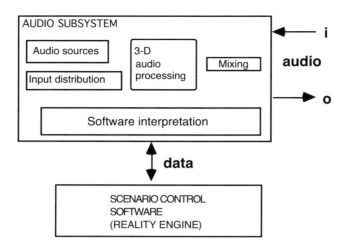

FIGURE 4.2. General system configuration for a **integrated** system. In this case, each of the subsystems exists within the computer. It is possible to assign one or more of these functions to an external device (the distributed approach).

computer mail) to building a virtual world in which a very detailed description of the acoustical environment is presented. The audio subsystem can be divided into hardware for **signal sources** and **signal processors** that are primarily dedicated to 3-D sound.

The audio subsystem is shown in greater detail in Figure 4.2 as a function of its more specialized components. The overall function of the audio subsystem is to manage internally the input and output of audio and control data to other audio components; to trigger signal sources from signal generators; to apply signal processing to various audio sources (viz., for spatial simulation and for speech recognition); to distribute and route internally generated or externally received audio source signals to one of any number of components; and to combine and mix signals for combining multiple outputs to one set of left-right outputs for each listener.

The audio subsystem responds to software calls from the **scenario control system (reality engine)** software. This system typically manages the input and output of multiple-function effectors and sensors that affect both the visual and audio simulations. The scenario control system can include data from a head position tracker or the localization of a mouse, for example. Data flows in the form of calls to and from the scenario control software to the audio subsystem software interpretation component. Here, software decodes calls from the reality engine into more complicated sets of instructions meaningful to the individual components of the audio subsystem.

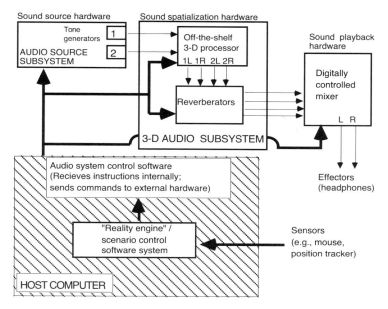

FIGURE 4.3. Detail of a **distributed** system using external devices for tone generation and spatialization. Data control signals are shown by the thick lines, while audio is passed via the thin lines. In this example, the 3-D audio subsystem is separate from the main computer. The scenario control system sends program calls to internal audio system software, which in turn sends appropriate control signals to the external audio hardware (e.g., via MIDI, described in Chapter 5). Note that the off-the-shelf 3-D sound system may include a separate computer. Four audio paths (1L, IR, 2L and 2R) result from the two inputs to the 3-D audio processor to provide separate left and right channels. The multiple left and right channels are combined for output to the headphones.

For example, consider a scenario involving a sink in the kitchen of a virtual house, written in the C programming language. The scenario control software, contained within the `main()` function, may use a call to a separate function called `play_water_sound(x,y,z)`, which is activated when the user turns the virtual sink handle. The relative location of the sink to the listener is passed to the audio subsystem software interpreter within the variables (x, y, z). In response, the software interpreter sends lower-level commands internally within the audio subsystem to switch on the recording of the water sound; send the location of the sink to the 3-D sound component; link the audio output of the signal source to the correct 3-D sound signal processor; and **mix** (blend) the audio outputs with any other sounds that occur simultaneously.

The routing of all of this information—both audio and data—can be accomplished on any number of levels of technical sophistication. At one extreme, the entire audio system just described can be contained within a single specialized chip, which can route sources between subsystems as a function of software commands. At the other extreme, one physically connects and disconnects cables between a set of audio devices (see Figure 4.3). More often than not, one ends up using a hybrid systems consisting of external (distributed) and internal (integrated) sound system elements.

Distributed versus Integrated Systems

Two approaches to implementing an audio subsystem in a virtual reality or multimedia system have already been introduced. The first involves assembling distributed, off-the-shelf components, either those intended directly for spatial manipulation and/or components such as synthesizers and effects processors. The second approach is to design an integrated system, consisting of hardware and software contained within the same computing platform as the scenario system itself. A hybrid approach, comprised of both integrated and distributed elements, is frequently used for practical reasons.

Distributed, off-the-shelf components normally are stand-alone systems. Configuration of these devices is usually cheaper than developing one's own components from scratch, especially if an off-the-shelf device is already well matched to the perceptual requirements. To the degree that the equipment can match the specific application and be adapted to specific individuals or applications, the more successful the distributed approach will be.

Perhaps someday you'll be able to purchase a 3-D sound system at the local electronics store that readily matches *any* system requirement. It will automatically account for the user's outer ear measurements and adjust to environmental noise conditions; sources will appear at a desired virtual location independent of head movement, with either headphone our loudspeaker playback. But currently available virtual acoustic devices have only partially satisfied one or more of these demands with their off-the-shelf products. Interestingly, there is an increasing trend for the development of "special purpose" devices for manipulating one aspect of spatial sound for a particular application rather than "one system does all, general purpose" machines (see Chapter 5). And, of course, there is always the lowest common denominator aspect of a mass-produced item, and the fact that costs are reduced (and the number of units sold increased) by compromising the sophistication of the human interface and of the device itself.

An assessment of how to implement virtual acoustics into a larger system should occur at the beginning of the design process and be translated into engineering requirements as soon as possible. The more that the audio system can be integrated within the hardware and software of the larger system, the better. This integration can mean involvement of additional personnel and expense: particularly, specialists in digital signal processing, firmware-hardware design, and digital audio production. But greater expenditure in the design phase allows the audio subsystem to match both the perceptual simulation and sound system integration requirements simultaneously. Another outcome of eliminating distributed devices is that the audio system becomes invisible to the user, and therefore is identified by the potential customer as an easily replicable system. Many a prototype demonstration has been received coldly because it appeared to be something imagined by either Rube Goldberg or Dr. Frankenstein. The bottom line is: integrate spatial audio processing and sound source system expertise as early as possible in the design of larger systems and use the distributed approach only when an integrated approach is not feasible.

Perceptual Requirements

Once the system requirements are known, the **perceptual requirements** of the 3-D sound system can be assessed. In Chapters 2 and 3, the diversity of human spatial hearing was overviewed with an emphasis on psychoacoustic findings. The ability to predict what a listener will hear—i.e., the match between source and listener by control over the medium—may matter or not in a specific application. Any assessment of perceptual requirements will need to consider the issues listed below.

- Two important cues for manipulating a sound source's lateral position are (1) interaural intensity differences and (2) interaural time differences. Over headphones, these cues are effective in themselves for lateralization— moving a sound source along a line inside the head.
- The spectral changes caused by the HRTF are the cues that help externalize sound images, disambiguate front from rear, and impart elevation information; the inclusion of the binaural HRTF is fundamental for a 3-D sound system that relates natural and synthetic spatial hearing.
- Changing HRTF spectra as a function of head movement is generally considered to improve overall localization performance. It does not necessarily follow that virtual source movement will improve performance.
- Nonindividualized HRTFs result in poorer localization than individualized HRTFs, but the latter are usually impractical for use in most applications. Under optimal conditions—individualized HRTFs, broadband signal, and target positions at lower elevations (eye level and below)—the localization error for virtual images is only slightly greater than under normal listening conditions. There are also notable localization performance differences between individuals.
- The perception of virtual sound sources is significantly affected by the inclusion of reverberation. Reverberation enables externalization of images outside the head, allows a sense of the surrounding environmental context, and is a cue to distance when used in terms of the R/D ratio. "Acoustically

correct" models of real or virtual environmental contexts can be computationally intensive and difficult to implement, but synthetic or first-degree approximations are possible. Reverberant sound also can cause a deterioration in localization accuracy of azimuth and elevation.

- Reversals of virtual images, particularly front-back, occur for a significant proportion of the population. The problem occurs more often with nonindividualized HRTFs than with individualized HRTFs, and more with speech than with broadband noise. However, the problem can be offset by previous expectation or associated visual cues, such as in a video game. Head movement cues should also help minimize reversals.

- Distance perception is not particularly good even under natural spatial hearing conditions. A decision needs to made as to what cues to implement—loudness/intensity, R/D ratio, and/or spectral changes due to air absorption. Intensity is the strongest cue; a decision needs to made as to whether it is implemented in terms of the inverse square law or a perceptual scale (such as sones). Cognitive cues (memory sand expectation, based on the type of sound) likely will override other cues. In addition, distance can be implemented as an absolute cue or as a relative cue.

DSPs FOR 3-D SOUND SIMULATION

Overview

Within a virtual reality or multimedia application, 3-D sound processing is almost always best accomplished digitally, using audio DSP chips to manipulate signal sources. A DSP can be used to process any type of digitally stored information: visual, audio, subsonic vibration, or magnetic fields. For example, NASA's SETI (Search for Extraterrestrial Intelligence) project analyzes faint "radio" signals from distant galaxies to explore the existence of other life forms. To extract the faint microwave signals received by radio telescopes, a specialized supercomputer for DSP operations is used, enabling 30 billion floating-point numerical operations per second. In audio, we can get by with specialized chip circuitry rather than a supercomputer, and thus far fewer mathematical operations. Audio DSP chips typically are designed to perform a limited set of digital DSP instructions *very quickly*—i.e., in real time. The number of instructions that can be accomplished is a function of the processor's **clock rate**, which coordinates tens of millions of instructions per second. Subsequently, the term "DSP" will be used to refer specifically to an audio DSP chip.

One can select a DSP that is most appropriate for a specific operation or type of digital filtering. Some DSPs can be programmed by using a limited but

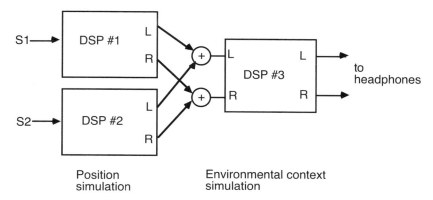

FIGURE 4.4. Separation of DSP sections with the 3-D audio component of the audio subsystem. S1 and S2 refer to two different sound sources to be spatialized by two different "position simulation" DSPs. DSP #1 and #2 each take a single sound source input and produce a binaural, two-channel output, simulating the signal at the outer ears in anechoic space. DSP # 3 is used for binaural environmental context simulation.

powerful **instruction set** to determine their operation. For digital filtering, these commands can center on a few mundane operations: e.g., obtaining a sound sample, multiplying it by another number in memory, adding it with another number, and placing the result at the output. Other DSPs are optimized to perform a specific complex operation and require little or no programming.

The 3-D audio component of an integrated audio subsystem (Figure 4.2) can contain anywhere from one to as many DSPs as can be accommodated. Each is usually assigned a specific function. For the moment, consider the case where a 3-D sound system is used to simulate the azimuth and elevation of two sound sources within a reverberant room (see Figure 4.4). A single DSP can perform the equivalent filtering operation for both the left and right pinnae, for simulating a single azimuth and elevation position. In this example, two DSPs are used to spatialize two sources to two different positions. Because both sound sources are in one room, the outputs of both positional DSPs can be fed to a common DSP that simulates stereo reverberation in a simple way. This is an example of a functional division between position simulation DSPs and environmental context simulation (reverberation) DSPs. Alternatively, a hybrid system approach can be used: the position simulation DSPs are part of the integrated system and the environmental context simulation DSP is within an off-the-shelf reverberator.

Thumbnail Sketch of DSP Theory

The following discussion gives a rudimentary, mostly nonmathematical overview of the theory behind DSP algorithms used for implementation of 3-D sound, particularly for implementation within an integrated system. This overview should enable the reader to conceptualize an overall system design for a particular application. The interested reader is referred to excellent introductions by Moore (1990) and Smith (1985), or more detailed books such as Oppenheim and Schafer (1975, 1989) to obtain a more in-depth discussion of DSP and digital filters. Support materials on audio applications that are available for specific DSP chips can be very helpful as well (e.g., the numerous implementation sheets for the Motorola 56001).

Impulse Response of a System

A **continuous, analog waveform**, like the sine wave shown in Figure 1.10, can be expressed as a function of continuous time, $x(t)$. Technically speaking, the symbols $x(t)$ indicate nothing about the waveform itself; we would need to know what x was equal to at each moment in time t. If $x(t)$ were a sine wave function with a particular frequency and phase, then indicating the variables in terms of $x(t) = \text{sine}(2\Pi \cdot \text{frequency} \cdot \text{time} + \text{phase offset})$ would be sufficient.

Unlike analog sound systems, where a signal $x(t)$ is transmitted via a continuously varying voltage, a digital system samples sound by obtaining a number from an analog-digital converter at discretely sampled intervals (e.g., in Figure 1.14, the sampled version of the sine wave of Figure 1.10). In other words, $x(n)$ is used to describe an audio input $x(t)$ sampled at discrete intervals of unit time (i.e., at the sampling rate).

The DSP effects a modification on the input data $x(n)$ such that the output $y(n)$ differs in some way. The internal operation performed by the DSP can be analyzed in terms of its **impulse response**, symbolized by $h(n)$ as shown in Figure 4.5. To make an unknown system $h(n)$ as shown in Figure 4.5 "identify itself," some kind of input is required. It's possible to identify a

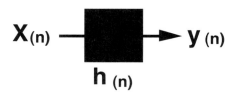

FIGURE 4.5. A generic signal processing algorithm. Here (n) refers to the time index of the digitized (sampled) waveform; x(n) is the input, y(n) is the output, and h(n) refers to a sequence that causes y(n) to differ from x(n).

Analytic impulse

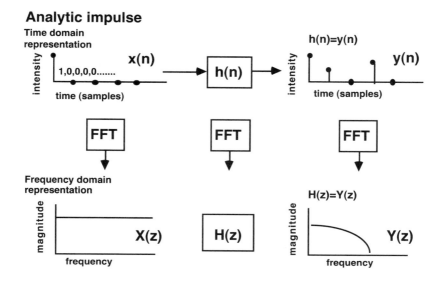

FIGURE 4.6. Use of analytic signal (here, a digital impulse) to determine the characteristics of a DSP system h(n). The time domain response of h(n) can be known via y(n) when x(n)= 1. If an FFT is applied to the time domain output, the spectral modification (transfer function) H(z) of the filter h(n) can also be determined from Y(z) since X(z) has a flat spectrum.

DSP's $h(n)$ by examining the digital output signal $y(n)$ after applying a specialized signal to the input known as **the analytic impulse**. The analytic impulse is a single 1 followed by a stream of 0s; its advantage is that $h(n)$ can be revealed by examining the transformation of this known input at the output $y(n)$. If $h(n)$ does not vary over time, then $y(n)$ will always equal $h(n)$.

This reveals the operation of the DSP in the **time domain**. But for digital filtering, it's more desirable to examine the DSP's effect in the **frequency domain**. To refer to the frequency domain (spectral content) of a time domain stream $x(n)$, the symbol $X(z)$ is used. Similarly, $Y(z)$ refers to the spectral content of $y(n)$, and $H(z)$ refers to the spectral content of $h(n)$. Not only is the time domain of the analytic impulse known in advance, but also its spectral content $X(z)$ is known; the FFT of the analytic impulse has a linear (flat) frequency response. (The term FFT is used throughout this discussion as shorthand to refer to both the discrete Fourier transform and the fast Fourier transform).

Figure 4.6 shows the time and frequency representations of the impulse, before and after processing by $h(n)$. By examining $Y(z)$—i.e., the FFT of $y(n)$—the **spectral modification** (or more properly, the **z transform**) of $h(n)$ can be described, because $Y(z)=H(z)$. This application of the FFT for spectral content and phase analysis is basically the same as shown for an arbitrary waveform in Chapter 1.

Simple DSP Systems

In audio, DSP algorithms are usually complex combinations of simple elements, which include some of the examples shown in Figures 4.7–9. The most basic operations are shown, which modify the input signal $x(n)$ by either a multiplication with a single value (Figure 4.7), a delay (Figure 4.8), or by addition with another signal (Figure 4.9).

The simplest way to view a digital multiply operation is as a virtual volume knob that can be applied almost anywhere in the digital circuit to an isolated stream of data (sound). For example, if the value of g in Figure 4.7 were set to 2, an input sequence at $x(n)$ of 1, 2, 3 would become 2, 4, 6 at $y(n)$. This is in fact a DSP system that increases gain by 6 dB. Following the internal workings of the DSP, the program memory would describe an operation to fetch a sample from the sequence $x(n)$, perform the multiplication operation $x(n) \cdot g$, and output the result at $y(n)$; the data memory would contain the value for g, and $h(n)$ would be the operation of multiplying g by the current value of $x(n)$.

Figure 4.8 shows a DSP system that performs a digital **delay** operation. A delay operation holds the sample value for a time indicated (in samples) in a digital **buffer**, a temporary storage area in memory that allows placing $x(n)$ values at one point in time and recalling them at another. Buffer sizes are limited by the size of the DSP's program memory. In Figure 4.8, a system that delays the output in relation to the input by one sample is shown; the sequence 1, 2, 3 applied to the system at $x(n)$ would result in the sequence 0, 1, 2, 3 at $y(n)$. If you want to build a DSP with a one-second delay, you could do it with a memory buffer with a length equal to the sample rate. If the sample rate were equal to 50,000, then the impulse response $h(n)$ would be 50,000 zeroes followed by a 1.

Finally, there's the digital summation operation, schematized in Figure 4.9. This system simply adds the value of a number of input signals, $x(n)1 - x(n)N$. A simple example of this for analog signals would be a stereo system Y cable used in for sending two outputs to one input. An implementation representative of the basic concept of a **digital mixing console** would include a gain factor applied to each input before summation. The implementation shown in Figure 4.9 shows a DSP system that combines two signals, $x(n)1$ and $x(n)2$, combined with a digital multiply g applied to $x(n)2$. So, if $g = 0.5$, and the sequence 2, 4, 0 were input at both $x(n)1$ and $x(n)2$, the output stream at $y(n)$ would be 3, 6, 0. Gain reduction is important in any digital mixing operation so that the summed signals do not exceed the maximum quantization value. Otherwise, **clipping** of the resultant waveform will result, which is audible as **distortion.**

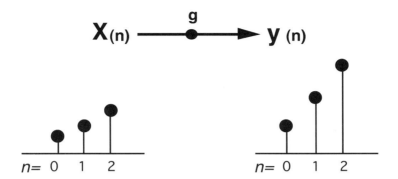

FIGURE 4.7. Top: multiply g that scales the value of x(n) at the output at y(n). Bottom: representation of the input sequence 1, 2, 3 and the output 2, 4, 6 when g is set to 2.

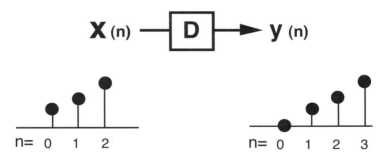

FIGURE 4.8. Top: a digital delay, where D refers to a delay of a given number of samples. Bottom: representation of the input sequence 1, 2, 3 and the output 0, 1, 2, 3 when D is set to 1.

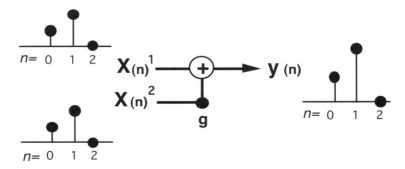

FIGURE 4.9. A summation of two sequences, x(n) 1 and x(n) 2, with a gain factor g = 0.5 applied to x(n) 2.

Implementation of Lateralized Positions

The gain and/or delay systems shown in Figures 4.7 and 4.8 can be used to impose simple lateralization cues on an input source. Figure 4.10 shows the use of a simple gain system for manipulating IID by adjusting the **interaural level difference (ILD)** in dB; Table 4.1 shows a scheme for calculating $g1$ and $g2$. Figure 4.11 shows a system using a delay for manipulating ITD.

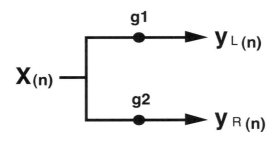

FIGURE 4.10. Two channel gain scheme (top). Refer to Table 4.1 for example values of g1 and g2.

g 1	g 2 (1 –g 1)	g1 (dB)	g2 (dB)	ILD (dB)
0.5	0.5	−6	−6	0
0.7	0.3	−3	−10.5	−7.5
0.8	0.2	−1.9	−13.9	−12

TABLE 4.1. Relationship between values for g1, g2 (1 − g1), their equivalent dB values (20 · log10), and the resultant ILD (interaural level difference). The value of 0.8 for g1 is appropriate for maximum displacement of the image, given the data shown in Figure 2.3.

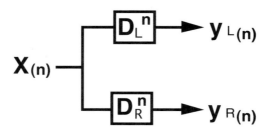

FIGURE 4.11. DSP system for producing ITD. As with the values shown for ILD in Table 4.1, the exact values for D can be calculated from a data set such as that shown in Figure 2.3. For example, to move a sound source to the extreme left side of the interaural axis, set the left delay to 0, and set the delay for the right channel to a number of samples equal to approximately one millisecond (see Figure 2.3).

Digital Filtering and Convolution

The basis of 3-D sound processing lies in imitation of the spatial cues present in natural spatial hearing, where the binaural HRTF imposes spatially dependent spectral modifications and time delays on an incoming wavefront to the hearing system. In virtual spatial hearing, it is possible to imitate these modifications using a DSP as a **digital filter.** One can obtain HRTF data by several methods (described later) and then use digital filters—specifically, two different $h(n)$s, termed a **binaural impulse response**—to produce a desired virtual acoustic source position. But before turning to this process, it's necessary to cover the basics of digital filtering.

What filtering does is **multiply** the **spectra** of two waveforms, which is equivalent to the **convolution** of the time domain representations of the waveforms. Stated mathematically, $X(z) \bullet H(z)$ in the frequency domain is equivalent to $x(n) * h(n)$ in the time domain (where * symbolizes convolution). This is exactly what the HRTF in natural spatial hearing does; in effect, the spectra of the pinnae is multiplied with the spectra of the sound source.

Convolution can be thought of as a time-indexed multiplication and summation operation performed on two numerical arrays, as shown in the four steps of Figure 4.12. The length of the output array $y(n)$ will be the length of the $x(n)$ and $h(n)$ arrays minus one. Assume that $x(n)$ is the waveform to be filtered, and $h(n)$ is the filter. Step 1 of Figure 4.12 shows $x(n)$ and $h(n)$ sequences that are to be convolved. Step 2 shows the entire $x(n)$ sequence multiplied with $h(0)$, the first coefficient in the $h(n)$ array, at time $(n) = 0$. Step 3 shows $x(n)$ multiplied with $h(1)$, the second coefficient in the $h(n)$ array, at time $(n) = 1$. Step 4 shows the resulting output $y(n)$, which is the summation of the values obtained in steps 2 and 3 at each sample interval. This process can be directly implemented into a DSP algorithm for filtering real-time signals by using techniques such as **overlap-add** and **overlap-save** methods (see Oppenheim and Schafer, 1975, 1989; Moore, 1990).

The two values used for the convolution operation $h(n)$ shown in Figure 4.12 are essentially two delays (with D equal to 0 and 1 sample) with each associated with a coefficient. The convolution operation can be represented by the DSP

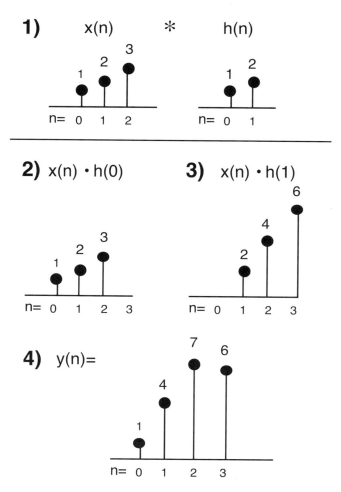

FIGURE 4.12. Convolution of two sequences in the time domain.

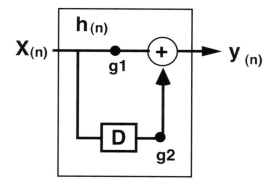

FIGURE 4.13. Configuration for a feed forward, finite impulse response (FIR) filter, y(n) = g1 • x(n) + g2 • x(n-1).

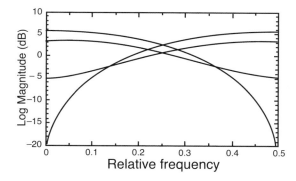

FIGURE 4.14. Transfer functions of the FIR filter shown in Figure 4.13 with the gain coefficient g1 set to 1, and g2 set to 0.9, 0.45, (the low-pass filters) and −0.45 and −0.9 (the high-pass filters). The simple transfer functions shown can become more complex when additional delay-gain systems are added to the filter structure; these can be tailored for various spectral modification purposes (e.g. for simulating walls, air attenuation, and HRTFs).

system shown in Figure 4.13. The DSP adds a scaled version of the sample currently at the input to a scaled version of the previously input sample, i.e., the one-sample delay. (The operation of delaying by a time value of 0 samples is not shown). The operation of outputting a sum of present and past inputs is referred to as a **finite impulse response (FIR) filter**, so-called because the length of $y(n)$ can be predicted from $x(n)$ and $h(n)$ as discussed previously; the result of applying an analytic impulse will be that $y(n) = h(n)$. Figure 4.14 shows the FFT of the impulse response of the system in Figure 4.13, using various positive and negative values for $g2$. A more complex transfer function can be achieved by convolution with additional coefficients, i.e. by adding additional delay and gain stages. For instance, to simulate a HRTF, up to 512 delay and gains may be used. Each delay-gain combination (including $D = 0$) is referred to as a **filter tap.**

What if, instead of summing the current input with a past input value, we were to use a **feedback loop** and sum the input with a delayed version of the *output*? In other words, the DSP memory buffer is filled with copies of $y(n)$ rather than $x(n)$, and then are summed at the output with the current input sample in delayed and scaled form. This is termed an **infinite impulse response (IIR) filter,** shown in Figure 4.15. This filter is "infinite" rather than finite due to the feedback loop; each successive output is made up of the input and scaled versions of previous *outputs*. For example, if the analytic impulse $(x(n) = 1)$ was applied to an IIR filter with $g1 = 1$ and $g2 = 0.5$, the output sequence would be 1, 0.5, 0.25, 0.125, .0625, etc., until the lowest sample value was reached. The value of $g2$ needs to be less than 1, otherwise the amplitude would increase to infinity with each output sample, eventually creating a situation that will make "pom-poms out of your speakers" (Smith, 1985). Figure 4.16 shows how different values for $g2$ allow the formation of simple low-pass and high-pass filters.

As will be seen later, IIR filters are useful in the simulation of reverberation, since the feedback operation is analogous to multiple reflections of sound in a room. And while binaural impulse responses cannot be implemented directly within an IIR filter, it is possible to imitate an HRTF in the frequency domain by using enough delay-gain sections, or combinations of FIR and IIR techniques. It is beyond the scope of this overview to cover what is known as a **biquad** configuration of a filter. This can be thought of as a combination of

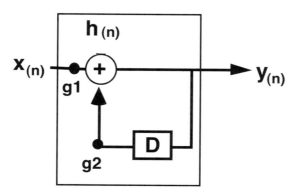

FIGURE 4.15. Configuration for a feedback, infinite impulse response filter, y(n) = g1 • x(n) + g2 • y(n–1).

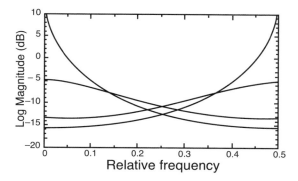

FIGURE 4.16. Transfer functions of the IIR filter shown in Figure 4.15 with gain coefficient g2 adjusted to 0.9, 0.45, −0.45 and −0.9, and gain coefficient g1 set to 0.9.

FIR and IIR systems, where the output is comprised of delayed, scaled versions of both the input *and* the output. Nevertheless, it is useful to note that more complex spectral shapings with arbitrarily-located peaks and notches can be obtained through the use of biquad filters, compared to the simple low- and high-pass transfer functions shown in Figures 4.14 and 4.16.

IMPLEMENTING HRTF CUES

Measurement of Binaural Impulse Responses

One can obtain the HRTF of an individual or of a dummy head by playing an analytic signal at a desired position at least a meter distant, and measuring the impulse response with **probe microphones** placed in the vicinity of the ear canals. This measurement then can be formatted for use directly within spatial filtering DSPs. Figure 4.17 shows the general procedures involving the **measurement, storage,** and **recall for simulation** of HRTF data. A sound source (e.g., a loudspeaker) generates an acoustical signal whose time and frequency characteristics are known in advance. Frequently, an acoustic version of the analytic impulse discussed in Figure 4.11 is used. An anechoic chamber is an optimal measurement environment because it minimizes the influence of reflected energy.

The **binaural impulse response** of the outer ear then is obtained from the pressure responses of the microphones and digitally recorded in stereo as a series of sampled waveform points. Both ears should be measured simultaneously; Figure 4.17 shows one microphone for clarity. An example of an impulse

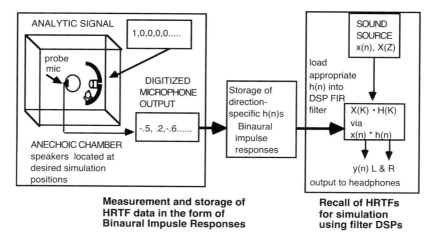

Measurement and storage of HRTF data in the form of Binaural Impulse Responses

Recall of HRTFs for simulation using filter DSPs

FIGURE 4.17. Overall plan of the HRTF measurement-storage-simulation technique. HRTFs are obtained using acoustic signals from loudspeakers that are recorded with probe microphones in the vicinity of the ear canals. The data are stored for input to DSP filters during the simulation stage. Usually, both ears are measured simultaneously. Not shown are the translation and possible compression of the measurements for use in the DSP. Note that each stage of measurement and simulation possibly can contribute significant influences to the overall character of the output signal, to the degree that each element is nonlinear (deviates from a flat frequency and phase response). In practice, some of these nonlinearities are difficult to control, although it is not certain how important they are psychoacoustically within a particular application.

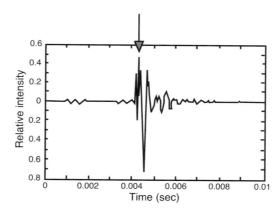

FIGURE 4.18. A time domain **impulse response** measured in the ear canal in an anechoic chamber. Its FFT would be equivalent to the HRTFs shown in Figures 2.13–2.15. The arrow indicates the peak of the response.

response recorded at one ear is shown in Figure 4.18. One performs this measurement operation for each desired virtual position and digitally stores the time domain signal from the left and right ears. The data for $y(n)L$ then is reformatted and loaded on demand into a DSP filter for left ear spectral modification, and similarly for the right ear. The sound source to be spatialized is filtered by these DSPs during simulation. Put another way, to multiply the spectrum of an input signal by the spectra of the outer ears, two separate time domain convolution processes are used for digital filtering of the input. Frequently, the impulse responses are modified in the time and frequency domain before formatting, as discussed subsequently.

Analytic signals can be used that have advantages over the impulse response method in terms of the signal-to-noise ratio and the ability to remove the spectral effects of early reflections in the measuring environment. One commonly used method is known as the **maximum-length sequence** (**MLS**) (Rife and Vanderkooy, 1989). This can be thought of as a stream of numbers with a known sequence in the time domain and, like the impulse, a theoretically flat spectrum. One can extract the impulse response of the system after using an MLS technique by **cross-correlating** the sequence with the resulting output, thereby **deconvolving** the pseudo-random sequence with the digital sequence recorded at the microphone. In other words, you get the same result as the impulse response, but with a better signal-to-noise ratio. The mathematical and computational implementation of these techniques can be found applied to HRTF measurements in Borish (1984) and Hammershøi *et al.* (1992); Rife and Vanderkooy (1989) give a comprehensive overview.

A technique related to MLS involves the use of **Golay sequences**, a particular random number sequence that when deconvolved is up to 30 dB less noisy than MLS sequences (Zhou, Green, and Middlebrooks, 1992). Another robust technique is to use a **swept sine** type of analytic signal; this technique is referred to as **time delay spectrometry** (**TDS**). The analytic signal in this case can be thought of as a sine wave that is swept quickly across the entire frequency range in a short period of time; like the MLS, the swept sine can be autocorrelated so as to obtain the equivalent of the impulse response. Many dual channel analysis systems, such as those manufactured by Crown, Brüel and Kjaer, and HEAD acoustics, feature this technique. Yet another way to analyze an HRTF in the frequency domain without specialized equipment is to play 1/3-octave noise bands from a high-quality loudspeaker and then measure the relative intensity and onset time using a digital waveform editor. One would then be faced with converting the 1/3-octave magnitude and phase analysis into a usable time domain representation for filtering.

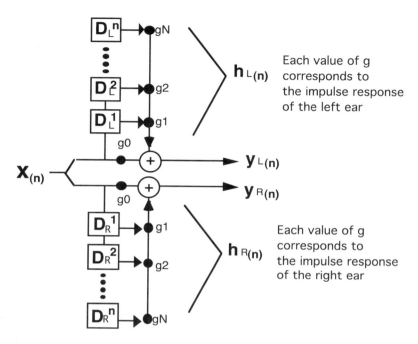

FIGURE 4.19. Convolution using two separate FIR filters for binaural output. Two impulse responses, one each for left and right ear HRTFs, are applied to the single input, resulting in a two-channel output.

Using Digital Filters for HRTFs

Setting up a DSP to filter a sound with an HRTF usually involves the use of an FIR filter with multiple delays and gains for convolution. This form of the FIR filter is also known as a **tap-delay line**, where "tap" refers to each of the *multiple,* summed delayed inputs that are placed at the output after scaling. As mentioned earlier, one can refer to a N-tap FIR filter; for instance, ten delay-gain stages would be called a ten-tap FIR filter. Each delay contains a previously input sample that is scaled by a particular coefficient, like the $h(n)$ section shown in Figure 4.13; but rather than using a single delay with a single coefficient for a gain, an inclusive set of delay-gains *1–N* are programmed into the DSP, and then summed for each output sample. The number of taps is typically 512 for a 50,000 kHz sampled response (duration = 0.01024 sec) to allow adequate time for both the temporal response of the pinna and the interaural delays. A smaller number of taps can be derived by one of the data reduction or synthetic HRTF techniques discussed below.

Figure 4.19 shows two tap-delay lines, one for each ear; note that this basically is a double expansion of the FIR filter shown in Figure 4.13. This system could be accomplished either with one DSP or with a separate DSP for each channel. The input $x(n)$ is sent to both tap delay lines. For each tap-delay line, the past N inputs are kept in a memory buffer and then summed together after multiplying with a value of g for each delay to produce an output sample. The gain values $g1$ through gN in the simplest implementation are loaded with a formatted version of the impulse response itself. The sound source to be spatialized is the input $x(n)$; the pinnae are considered the filters $h(n)L$ and $h(n)R$; and the outputs $y(n)L$ and $y(n)R$ result from the convolution of $x(n)$ with left and right $h(n)$s.

Collecting HRTF Measurements

No single scientifically accurate method has been established for measuring HRTFs. Researchers constantly are exploring new techniques and equipment in order to minimize measurement *variability* (i.e., the variation in successive measurements that should remain constant), to improve the signal-to-noise ratio of the measurement hardware used and to determine the optimal placement of the measurement microphone. Indeed, one can find a wide array of methods, ranging from scientific to exploratory to fantastical, by comparing the scientific literature, binaural recording practices, and methods outlined in U.S. patents related to 3-D sound (for instance, the empirical method described in Lowe and Lees, 1991).

The 3-D sound system designer in most cases will want to obtain a library of HRTF measurements, from which particular sets can be selected on the basis of either user preference or behavioral data. Several corporate and university research projects are underway where HRTF measurements are collected and subsequently utilized for digital filtering. Someday, a collection may be available on a CD-ROM; already, some HRTFs are available over computer networks (see Chapter 6). But because the process is extremely time consuming and relatively difficult, most people who have gone to the trouble to do the work are unwilling to merely give away the data. A notable exception is the paper by Shaw and Vaillancourt (1985), essentially an archival report of HRTF measurements discussed in an earlier paper of Shaw (1974). This paper gives HRTF magnitude data in numerical form for 43 frequencies between 0.2–12 kHz, the average of 12 studies representing 100 different subjects. However, no phase data is included in the tables; group delay simulation would need to be included in order to account for ITD.

In 3-D sound applications intended for many users, we want might want to use HRTFs that represent the common features of a number of individuals. But another approach might be to use the features of a person who has desirable HRTFs, based on some criteria. (One can sense a future 3-D sound system where the pinnae of various famous musicians are simulated.) A set of HRTFs from a **good localizer** (discussed in Chapter 2) could be used if the criterion were localization performance. If the localization ability of the person is relatively accurate or more accurate than average, it might be reasonable to use these HRTF measurements for other individuals. The *Convolvotron* 3-D audio system (Wenzel, Wightman, and Foster, 1988) has used such sets particularly because elevation accuracy is affected negatively when listening through a bad localizer's ears (see Wenzel, et al., 1988). It is best when any single nonindividualized HRTF set is psychoacoustically validated using a statistical sample of the intended user population, as shown in Chapter 2. Otherwise, the use of one HRTF set over another is a purely subjective judgment based on criteria other than localization performance.

The technique used by Wightman and Kistler (1989a) exemplifies a laboratory-based HRTF measurement procedure where accuracy and replicability of results were deemed crucial. A comparison of their techniques with those described in Blauert (1983), Shaw (1974), Mehrgardt and Mellert (1977), Middlebrooks, Makous, and Green (1989), Griesinger (1990), and Hammershøi *et al.* (1992) is recommended for a technical overview of the differences in microphone placement and analytic signals used. Wightman and Kistler (1989a) placed the small probe microphone shown in Figure 4.20 in the ear canal, at a distance of approximately 1 to 2 millimeters from the eardrum. By doing this, the ear canal resonance could be included in each measurement taken. An extremely thin (<0.5 mm) custom-molded shell made of Lucite was used to mount the microphone in the entrance of the ear canal in such a way as to occlude the ear canal as little as possible. Figure 4.21 shows the probe microphone as placed in the ear for an HRTF measuring session.

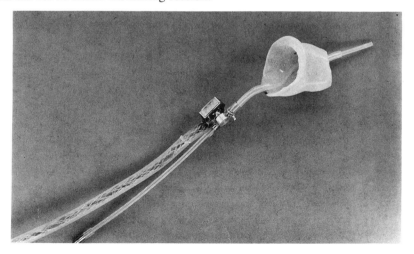

FIGURE 4.20. Probe microphone used for HRTF measurements near the eardrum, by Wightman and Kistler (1989a). *Courtesy of Fred Wightman and Doris Kistler, University of Wisconsin, Madison; used by permission of the American Institute of Physics.*

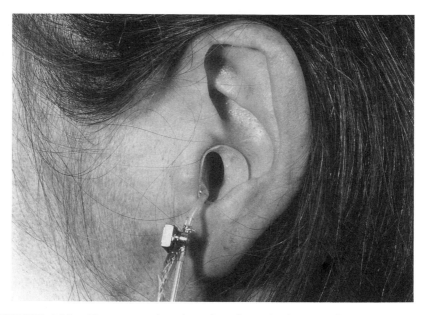

FIGURE 4.21. Placement of probe microphone in the ear. *Courtesy of Fred Wightman and Doris Kistler, University of Wisconsin, Madison; used by permission of the American Institute of Physics.*

Figure 4.22 shows a subject being measured in the Waisman Center anechoic chamber at the University of Wisconsin, Madison. Eight loudspeakers were mounted at ±36, ±18, 0, 54, 72, and 90 degrees elevation on a movable arc at a distance of about 1.4 meters from the head. This placement allowed all desired elevations to be obtained at a particular azimuth before realignment of the speaker arc. The equipment used included Realistic MINIMUS-7 speakers for playing the analytic signal; Etymotic miniature microphones with the custom shell; and compensatory equalization for Sennheiser HD-340 headphones.

An alternative method is that used by Hammershøi, *et al.* (1992), where HRTF measurements were taken at the *entrance* of a plugged ear canal. A single measurement for the ear canal resonance was used to postequalize the HRTF measurements. This technique was illustrated earlier in Chapter 2, Figure 2.11; recall that the ear canal transfer function for the most part is independent of direction. This method is easier to accomplish compared to the inside of the ear canal methods used by Wightman and Kistler (1989a), since a miniature microphone can be placed more easily than a probe mic, and a better frequency response and lower noise can be obtained (Hammershøi *et al.*, 1992 used a Sennheiser KE 4-211-2). On the other hand, the initial measurement of the ear canal resonance becomes critical, and one is still faced with mounting a tiny probe mic in the ear canal.

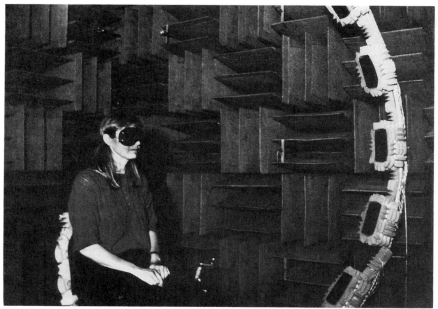

FIGURE 4.22. A subject in the anechoic chamber at the psychoacoustic laboratory at the Waisman Center of the University of Wisconsin, Madison. *Courtesy of Fred Wightman and Doris Kistler.*

To measure HRTFs from different directions, one can move the analytic signal source around listeners; have the subjects move their head to a calibrated position; or use a multiple speaker measurement system such as the Auditory Localization Facility shown in Figure 4.23, built at the Armstrong Laboratory of Wright-Patterson Air Force Base (McKinley, Ericson, and D'Angelo, 1994). Each of the 272 speakers are mounted in a 14-foot diameter geodesic sphere, with a speaker at every 15 degrees azimuth and elevation. The location of each speaker represents a measured position; the listener and the speakers do not need to be moved, facilitating the collection of the measurements. Note that the cage struts are covered with sound-absorbing material; this is necessary to help minimize reflections from the surfaces of the sphere so they are not included in the impulse response.

Considering that perceptual performance is optimized when listening through one's own ears, it would be ideal to design a system that quickly obtained the HRTFs of a particular user prior to virtual acoustic simulation. A prototype of such a system was developed by Crystal River Engineering, complete with a voice that instructs a user how to move their head and then stop for a measurement. The difficulties facing such a system again have to do with

FIGURE 4.23. The "speaker cage" developed at Armstrong Laboratory, Wright-Patterson Air Force Base for HRTF measurements. Note the sound absorbing foam around the struts. *Courtesy of Richard McKinley, Armstrong Laboratory, Wright-Patterson Air Force Base.*

variability in microphone placement and dealing with reflective environment information in the impulse response.

In a laboratory environment, one ideally can control for all nonlinear effects, including sound reflections, and for accuracy of placement of the person being measured, the microphone, and the sound source. But if one is willing to sacrifice replicability or scientific accuracy, the need for an anechoic environment and accurate compensation for the measurement system can be relaxed. The impulse response used in a filter need not only represent the HRTF itself; it can include environmental context information as well. Of course, this procedure is not acceptable if a more general approach is desired to simulate environmental contexts. If one starts with anechoic HRTFs, reverberation can be added at a later point, but its more difficult to go the other way around.

Figure 4.24 shows a composer making a binaural recording using the Sennheiser MKE-2002 stethoscopic microphones. Electret-condenser omnidirectional microphone capsules are positioned on the head with small hooks that go inside the ear; the mics sit out from the ear canal entrance about 1 to 3 mm. Tom Erbe and the author have conduced informal experiments in collecting "cheap and dirty" HRTFs using these microphones, with balloon pops as the analytic signal, a classroom and concert hall as the environmental contexts, and no postequalization. While the results were certainly by no means optimal or even acceptably noise-free, it was possible to convolve sound sources with these impulse responses using a desktop PC and capture a sense of the acoustic environment the original measurements were made in. Most likely the spectra of the actual HRTF was completely altered, but the ITD, IID, and reverberation cues were reasonably well retained. This preservation of powerful localization cues is the working premise of many commercial 3-D sound systems.

A less cheap and dirty approach is as follows. Given an anechoic chamber, insert microphones, patient subjects, and a fairly expensive set of test equipment, one can build ones own HRTF measurement laboratory. Hewlett Packard, HEAD acoustics, and Brüel and Kjaer among others make specialized dual-channel hardware for generating analytic signals and recording impulse response data (see Chapter 6). One can then custom tailor simulations by creating libraries of HRTFs with characteristic features that can be accessed by software routines to set up the DSP filters for a particular user.

Calculation of Generalized HRTF Sets

In many applications, the taking of individual HRTF impulse response measurements is impractical; it would be better to use measurements that represent the best compromise between the HRTF features of a given population or the good localizer paradigm mentioned earlier. Chapter 2 reviewed psychoacoustic data on the differences between individualized and nonindividualized HRTF localization performance; anecdotally, many have reported that frontal imagery and elevation simulations work dramatically better when listening through one's own HRTF measurements. But for most 3-D sound applications some type of generalized HRTF set will be necessary.

FIGURE 4.24. Using the stethoscope microphones of the Sennheiser MKE-2002. Composer Chris Brown recording a fireworks show for his electronic music composition, "Year of the Rooster." The microphones are omnidirectional, and set more outside the head than with an insert microphone. The MKE-2002 is supplied with a mannequin head that can be used as an alternative to the stethoscope feature. *Photo courtesy Chris Brown; imaging by Denise Begault.*

Averaged HRTFs can originate from an analysis of either physical or spectral features. For instance, the size of the cavum conchae and the location of the entrance of the ear canal can be measured for a large number of people and then averaged. Another method would be to examine the spectra of many binaural impulse responses via their Fourier transforms and perform a "spectral averaging," similar to the work of Shaw (1974) discussed earlier. It is possible that this type of averaging will cause the resultant HRTF to have diminished spectral features relative to any particular individual's spectral features. In the extreme case, one person has a 20 dB spectral notch at 8 kHz, and another has a 20 dB spectral peak—the average is no spectral feature at all.

Another method is to use a technique termed **principal components analysis (PCA)**. This is a technique where all of the HRTFs for different positions are evaluated simultaneously; using statistical techniques, one can isolate spectral features that change as a function of direction and those that remain more constant (Martens, 1987; Kistler and Wightman, 1992). The result of PCA is a set of curves that fall out of correlation analysis and that collectively explain all of the variation in the analyzed data. For instance, Kistler and Wightman (1992) found five "basis functions" that could be selectively weighted in order to produce any particular HRTF. Their studies comparing localization performance with PCA-derived HRTFs and measured transfer functions suggest that the PCA versions are as robust as the actual measurements; a study with five subjects showed a high correlation between responses to the synthesized and measured conditions. However, when the accuracy of the synthesized HRTF was diminished by reducing the number of basis functions, an increase in front-back reversals occurred.

Different types of "average" binaural impulse responses can be collected by obtaining HRTF impulse responses from several dummy heads. Each dummy head represents a particular design approach to creating a "standardized head" with average pinnae. The procedure for measurement usually is easier, since many dummy heads are part of integrated measurement and analysis systems. The low frequency response will be better than with probe mics, since the mic is built into the head; the results will be more replicable since the mic and head remain fixed in position. Another advantage is that 3-D sound systems based on dummy head HRTFs will be closely matched to actual recordings made by the same binaural head, allowing compatibility between the two different types of processing.

When using dummy head HRTFs, one must assume the manufacturer has constructed features of the outer head and pinnae according to their advertised specifications and that they will indeed function well for localization of virtual sources. However, there are bound to be differences in timbre and spatial location that can be evaluated empirically. One head might sound more natural to a particular set of users than another. This discrepancy can be due to the microphones used, the manner used for simulating the ear canal, or how closely the dummy head's dimensions match those of the listeners. Another important reason has to do with the equalization of the head. For instance, the Neumann KU-100 is designed to be compatible with normal stereo loudspeaker playback or headphones. The AACHEN head allows switchable equalization between free field and "independent of direction" curves, where the average frequency response from all directions is "factored out" of the overall frequency response. The Neumann KU-100 is equalized for diffuse-field headphone playback, and the AACHEN system is equalized optimally for an integrated headphone playback system (equalization is discussed below in greater detail).

The size of the head (and correspondingly, the head diffraction effects and overall ITD) is probably a major component in the suitability of one dummy head versus another because of the importance of the interaural delay present in HRTFs for spatial hearing (Wightman and Kistler, 1992). For instance, because Asians on average have a smaller head size than average Europeans, and correspondingly smaller pinnae, two versions of the same dummy head are available from one manufacturer.

Some of the current choices for dummy heads for both measurement and recording are illustrated in Figures 4.25–4.29. Figure 4.25 shows the KEMAR, a standard audiological research mannequin manufactured by Knowles Electronics. A complete head and torso simulator, its design evolved through a careful averaging of anthropomorphic data (Burkhard and Sachs, 1975). It allows several varieties of pinnae to be mounted on the head for a variety of acoustic measurement applications; it also includes an ear canal simulator based on the work of Zwislocki (1962). A review of the literature shows that KEMAR's HRTFs have been used many times in both 3-D sound applications and localization studies (e.g., Kendall, Martens, and Wilde, 1990). The coupling of the ear canal simulator to internal microphones (typically, a laboratory microphone such as the Brüel and Kjaer 4314) is designed to meet international standards, facilitating audiological research for hearing aid devices.

Figure 4.26 shows the HATS (Head and Torso Simulator) developed by Brüel and Kjaer (model 4128), for acoustical research, evaluation of headphones and hearing protectors, and other types of binaural measurement; it includes a mouth simulator as well. The geometry of the head is symmetrical, based on average adult anthropometric data. Brüel and Kjaer also manufacture the Sound Quality Head and Torso Simulator (model 4100) which uses a free-field equalized system, "soft shoulders" to imitate clothing, and a smaller torso profile; it is more suitable for binaural recordings due to its "directivity optimized for sound image location." Both interface with a range of Brüel and Kjaer analysis systems useful for measuring HRTF impulse responses (e.g., their Dual Channel Analyzers, models 2032-2034).

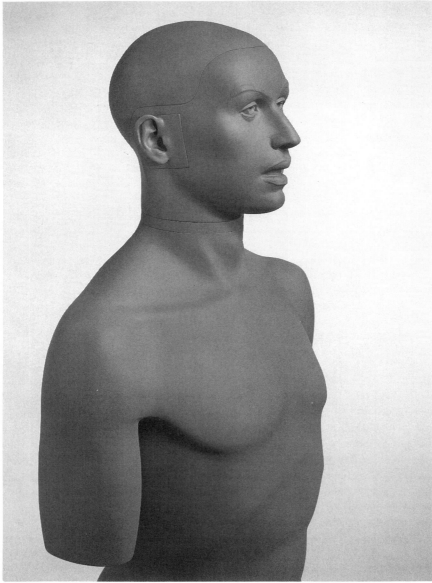

FIGURE 4.25. The KEMAR mannequin head and torso. KEMAR is used extensively in audiological research, particularly for hearing aids; several types of pinnae are available. Standard laboratory measurement microphones attach to a Zwislocki ear canal simulator inside the head. *Photograph courtesy of Knowles Electronics.*

FIGURE 4.26. The Brüel and Kjaer HATS head and torso simulator (model 4128) for acoustical research in areas such as evaluation of headphones, telephones, hearing aids and hearing protectors. The geometry of the head is symmetrical, based on average adult anthropometric data. *Photograph courtesy of Brüel and Kjaer.*

Figure 4.27 shows the AACHEN Head HMS II manufactured by HEAD acoustics. The outer ears of the head do not resemble the outer ears of a normal person; rather, the structures used capture averaged features of different pinnae. Note also that the upper part of the body and the torso are part of the design. The HMS II is part of an integrated measurement and analysis system, whereas the similarly structured HRS II is oriented toward binaural recording; the BAS (Binaural Analysis System) is optimized around the HMS II for many types of signal measurement and analysis. The design of the head is based primarily on the research of Klaus Genuit (1984); his approach was novel in that he analyzed the HRTF by decomposing the pinnae into physical models of simple structures such as disks and also investigated the effects of the torso and shoulders. These elements were combined into the "structural averaging" of the AACHEN Heads, resulting in optimized, averaged HRTFs that are at least as good as more anthropomorphic dummy heads.

Figures 4.28–4.29 show the KU-100 binaural head, known as "Fritz II," manufactured by Neumann GmbH. The design was refined from earlier models KU-80 (publicly introduced in 1973) and KU81 ("Fritz I"); the former was free-field equalized, while the KU-81 and KU-100 are equalized to a diffuse field. This equalization allows compatibility with loudspeaker playback and with traditional stereo micing techniques. Unlike the KU-81, the KU-100 has symmetrical left-right pinnae and placement; the ear coordinates were changed from the earlier model "thereby avoiding vertical displacements of sound sources (artificial elevations)" during playback (Neumann, 1992).

Equalization of HRTFs

Raw impulse measurements must undergo both time and frequency domain modifications before being formatted for use in a DSP. The time domain modifications are relatively straightforward, while the frequency domain modifications involve attempts to apply postequalization to the HRTF to eliminate errors in the initial measurement and are therefore more complicated. There will also be a need for numerical formatting and conversion that will be inherent to the particular architecture of the DSP used, e.g., converting from 16-bit floating point to 24-bit signed integers. There are usually software routines that readily perform these conversions within integrated measurement and synthesis systems or signal-processing packages.

The first time domain modification to be performed on a set of raw impulse responses is to discard the "blank" portion at the beginning that results from the time it takes the impulse to travel from the speaker to the microphone. This can be applied to all HRTFs by investigating the case involving the shortest path, i.e., the measurement position with the ear nearest the speaker. Once this is done, a **normalization** procedure can be applied to make best use of the

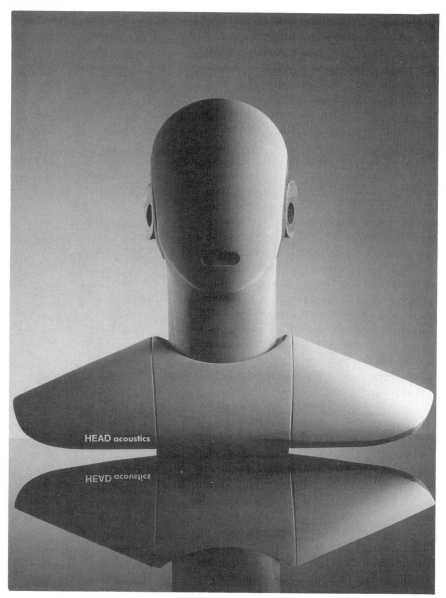

FIGURE 4.27. The AACHEN Head Model HMS II, manufactured by HEAD acoustics. The outer ears are designed according to a "structural averaging" of HRTFs; note the inclusion of the upper torso. Two versions of the head are available: the HMS II, intended for analysis within an integrated system, and the HRS II, for recording studio applications. *Photograph courtesy of Wade Bray, Sonic Perceptions–HEAD acoustics.*

FIGURE 4.28. The Neumann KU-100 dummy head. This is the third generation of Neumann binaural recording devices (following the KU-80 and KU-81). *Photograph courtesy of Juergen Wahl, Neumann USA.*

FIGURE 4.29. Inside of the Neumann KU-100 showing microphone capsules and power supply. *Photograph courtesy of Juergen Wahl, Neumann USA.*

available digital quantization range. A normalization procedure involves finding the loudest sample in a given sequence and then multiplying all samples in that sequence so that the loudest sample is at the maximum quantization value. The impulse response with the shortest blank portion at its beginning is usually the one with the loudest sample. The coefficient used to normalize the loudest impulse response is then applied to all impulse responses; note that if one normalized all impulse responses individually, IID cues would be compromised.

A final type of time domain modification can occur either at the measurement stage or can be placed at the interface in the form of user-interactive software. Because of the importance of overall ITD (Wightman and Kistler, 1992), one can customize the delay inherent to each binaural HRTF impulse response pair. Specifically, by inserting or subtracting blank samples at the start of the impulse response corresponding to the ear furthest from the analytic signal, the overall ITD cue can be customized to a particular head size, or even exaggerated.

Postequalization of HRTFs is desirable in order to eliminate the potentially degradative influences of each element in the measuring and playback chain. Most of the spectral nonlinearities originate from the loudspeaker used to play the analytic signal, the measuring microphone, and the headphones used for playback. A simplified example is as follows: If the frequency response of the loudspeaker used for the analytic signal is known to have a 6 dB dip at a center frequency of 500 Hz, compensatory postequalization should be applied to the HRTFs using a linear-phase filter with a 6 dB gain at the same center frequency. The transfer function of the probe microphone must usually be compensated for, since small microphones are often difficult to design with a linear frequency response. They are also very inefficient at low (< 400 Hz) frequencies, making high-pass filtering or "bass boosting" a fairly common HRTF postequalization procedure (see later).

A frequency curve approximating the ear canal resonance can also be applied for analysis purposes, if it is not part of the impulse response measurement. This can be done, albeit with great difficulty in the measurement phase, by postequalization with an actual measurement of the ear canal, as discussed earlier in connection with Hammershøi *et al.* (1992). Alternatively, an approximation of the ear canal resonance can be imparted using a standard equalization, such as that described in Zwislocki (1962). Since the ear canal resonance is generally considered to be the same for all angles of incidence this needs to be done only once. For most applications of 3-D sound, the listener's own ear canal resonance will be present during headphone listening; this requires *removal* of the ear canal resonance that may have been present in the original measurement to avoid a "double resonance." See Møller (1992) for an extensive discussion of frequency-domain equalization of HRTFs.

One of the most important aspects of effectors has to do with the relationship between the equalization of the headphones used for playback of 3-D sound and dummy head recordings. If the frequency response of the headphones used during playback is known, then compensatory equalization can be applied to the HRTFs to make the headphone frequency response transparent. One solution that bypasses many of the problems associated with normal headphones is the "tubephones" manufactured by Etymotic research (model ER2, flat within ±1.5 dB from 0.2 –10 kHz), but one must insert them into the ear canal at a precise distance to achieve the optimal frequency response curve. Most users of 3-D sound systems will use either **supraaural** (on the ear) or **circumaural** (around the ear) headsets. Circumaural headsets are used in professional designs since speaker diaphragms with better frequency responses can be used, greater isolation from extraneous noise can be achieved, and better, more consistent coupling between the ear and the headset is insured.

The issue of headphone equalization has been dealt with extensively in work conducted at the Institute of Broadcast Technique (IRT) in Munich toward an international standard (Theile, 1986). Until recently, almost all professional headphones were designed to have **free-field equalization**. Consider a listener in an anechoic chamber, located at a distance from a white noise sound source directly ahead at 0 degrees azimuth and elevation; a spectral measurement K is taken at the eardrums. A free-field equalized headphone playing a recording of the noise should produce the same spectrum K at the eardrum of the listener as in the anechoic chamber. In this case, reproduction is optimal for the 0 degree azimuth and elevation position—but is not optimal for virtual sound sources incident from any other direction. Due to incompatible spectral modification between the free-field curve and the HRTF from positions other than 0 degrees, IHL and "tone color defects" result with free-field equalized headphones (Theile, 1983, 1986).

Recall from Chapter 3 the concept of **critical distance,** defined as the location where the reverberant field is equal in level to the direct field. As one moves away from a virtual source toward and beyond the critical distance, the sound field increasingly will become random in its direction of incidence. In other words, the incoming sound field is *diffused* about the listener rather than from a specific direction, as in the free-field case. Since sound is heard via both direct and indirect paths, it makes more sense to consider all angles of sound source incidence when equalizing headphones. For this reason **diffuse field equalization** takes into account the level differences for "random" directional incidence of noise in third-octave bands. One obtains an "average HRTF" for multiple angles of sound incidence and then compensates for this curve in the headphone design by subtracting this response from the headphone free-field response. This equalization results in corrections on the order of ± 1.5–2.5 dB across frequency. Headphones that have a "flat" diffuse field equalization are available from several manufacturers and are well matched to dummy heads that are similarly equalized.

The best way to compensate for a headphone frequency response is to conduct measurements in tandem with the ear of a particular listener, since the headphone-ear coupling for an individual listener will differ in each case. This affects higher frequencies in particular, since every person "fits" a pair of headphones differently. In the work of Wightman and Kistler (1989a, 1989b), the HRTFs of the measured individuals were used at a later point in localization studies, using the same individuals as subjects. They localized actual loudspeakers and then listened through their own simulated pinnae and localized virtual loudspeakers. Thus the loudspeaker transfer function did not need to be compensated for. The ear canal transfer function was included in the HRTF measurements since the probe microphone was placed as near to the eardrum as possible. To make the actual source and virtual source conditions identical, they first took the headphone and ear canal transfer function for each subject. This was accomplished by transducing the same analytic signal as used in the HRTF measurement phase through headphones instead of the loudspeakers. Second, a filter with the inverse transfer function of this measurement was calculated to postequalize the obtained HRTFs. As a result, headphone playback of virtual stimuli was "neutral," i.e. equal to free-field listening.

This compensatory equalization can be thought of as a spectral *division* in the frequency domain. Commercially available signal-processing packages allow determination of these types of "inverse" filters, but the design is not always straightforward. Care must be taken with numerical overflow and unstable filter designs (see Oppenheim and Shafer, 1989). It is possible to describe this process theoretically using a simple mathematical representation, where the spectra of each element are either multiplied with or divided from each other. Note that when the spectrum of a particular element is in both the numerator and denominator of the equation they cancel each other out. The frequency domain transfer functions for one ear can be represented as follows:

$A(Z)$ = *analytic signal*;

$M(Z)$ = *probe microphone*;

$C(Z)$ = *ear canal*;

$L(Z)$ = *loudspeaker*;

$HP(Z)$ = *headphone*;

$H(Z)$ = *naturally occuring HRTF in a free field*;

$RAW(Z)$ = *uncorrected HRTF for virtual simulation*;

$COR(Z)$ = *corrected HRTF for virtual simulation*;

$INV(Z)$ = *inverse filter for correcting RAW(Z)*;

$YE(Z)$ = *the signal arriving at the eardrum*;

$YM(Z)$ = *the signal arriving at the probe mic*; and

$X(Z)$ = *an input signal to be spatialized*.

In natural spatial hearing, a sound source played by a loudspeaker can be described as

$$YE(Z)_{natural} = X(Z)H(Z)C(Z)L(Z).$$

In a virtual spatial hearing simulation, we desire

$$YE(Z)_{virtual} = YE(Z)_{natural}.$$

First, the uncorrected HRTF is measured for a particular direction by playing the analytic signal through the loudspeaker:

$$RAW(Z) = YM(Z) = A(Z)M(Z)C(Z)H(Z)L(Z).$$

Second, the headphone and ear canal are measured by playing the analytic signal through the headphones:

$$YM(Z) = A(Z)M(Z)C(Z)HP(Z).$$

To obtain $COR(Z)$, it is first necessary to find the inverse filter $INV(Z)$:

$$INV(Z) = \frac{1}{A(Z)M(Z)C(Z)HP(Z)}$$

Then the uncorrected HRTF is corrected as follows:

$$COR(Z) = RAW(Z)INV(Z)$$
$$= \frac{A(Z)M(Z)C(Z)H(Z)L(Z)}{A(Z)M(Z)C(Z)HP(Z)}$$
$$= \frac{H(Z)L(Z)}{HP(Z)}$$

To create a virtual sound source, the spectra of the input is convolved with $COR(Z)$, and then played through headphones via the ear canal of the listener:

$$YE(Z)_{virtual} = [X(Z)COR(Z)][HP(Z)C(Z)]$$
$$= \left[X(Z)\frac{H(Z)L(Z)}{HP(Z)} \right][HP(Z)C(Z)]$$
$$= X(Z)H(Z)C(Z)L(Z)$$

The final part of the previous equation shows that natural spatial hearing and the virtual simulation are equivalent:

$$YE(Z)_{natural} = YE(Z)_{virtual} = X(Z)H(Z)C(Z)L(Z).$$

Equalization of HRTF impulse responses also can be applied on the basis of preference within the intended application. From a commercial standpoint, one usually requires more low-frequency information than can be obtained from a

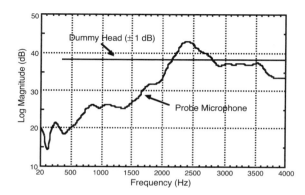

FIGURE 4.30. 20 Hz–4 kHz frequency response of an HRTF measured with a probe microphone, compared to the published frequency response for a dummy head (0 degree azimuth and elevation). The frequency region between 20–500 Hz is approximately 25 dB down relative to the peak at 2400 Hz. A chromatic scale played on a piano through the HRTF obtained with the probe microphone would sound very unnatural, especially on the lower half of the keyboard.

probe or insert microphone. Figure 4.30 shows a comparison of a 0 degree elevation and azimuth HRTF measured with a probe microphone versus one from a dummy head. Note that the probe microphone frequency response falls off at frequencies below 2.5 kHz, making this HRTF difficult to use for music recording applications (Begault, 1991b). A solution adopted by some has been to include a bass boost to the lower frequency region by creating new, spectrally altered versions of the measured HRTFs.

Data Reduction of HRTFs

Methods for Data Reduction

For hardware implementation purposes, it is often desirable to create versions of HRTFs with shorter impulse response lengths. For example, a Motorola 56001, used for stereo FIR filtering at a sampling rate of 44.1 kHz, can accommodate a maximum of 153 coefficients in the impulse response with a 27 MHz clock rate (Burgess, 1992), whereas 512–1024 coefficients are usually obtained in a raw impulse response measurement at the same sampling rate. There are three basic methods to convert a raw HRTF impulse response into a version with fewer coefficients. The first is to simply convert the sample rate of the measured impulse response to a lower value. Many HRTF measurements contain substantial amounts of noise at frequencies above 15 kHz, making **downsampling** a viable option. The problem is that most of the audio world is tied to a 44.1 kHz sampling rate; one would need to allocate DSP resources towards converting input signals to the same sampling rate as the resulting filter.

The second method is to perform a simple rectangular **windowing** of the impulse response. The simplest way to reduce HRTF data is to eliminate lower-amplitude portions at the start and end of the impulse responses, keeping only the higher-amplitude portions. A rectangular window of a frame size smaller than the original measurement is applied to capture the essential features of the HRTF, as seen in Figure 4.31. The window needs to be applied at the same point for all measured HRTFs to preserve the ITD information. A potential problem with this technique is the loss of low-frequency information, if indeed these frequencies were actually captured in the initial measurement. One commercially available 3-D sound system used this technique in its original software implementation to allocate filtering resources between one and four sources: one source would be filtered by 512 taps, two by 256, and four by 128 points each, based on a windowing system similar to that seen in Figure 4.31.

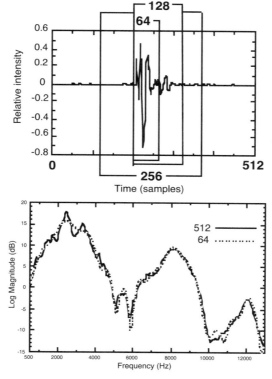

FIGURE 4.31. Time domain windowing (top) of an original impulse of 512 points narrowed to 256, 128, and 64 points; (bottom) a comparison of the magnitude response of the original impulse and a 64-point windowed version.

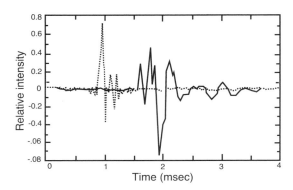

FIGURE 4.32. Minimum-phase version (dotted line) of a measured impulse response measurement. The process requires "reinserting" an approximation of the delay.

The third method is to use statistical or other analytic means to examine measured HRTF magnitude responses and then produce shorter impulse response length filters that approximate the original HRTFs. This technique results in **synthetic HRTFs** and synthetic binaural impulse responses. The synthetic HRTF method involves taking a measured HRTF and producing data that meets both perceptual and engineering criteria: viz., a synthetic HRTF should sound the same as a measured one, but be more amenable to a signal-processing algorithm. Studies by Asano, Suzuki, and Sone (1990), Kistler and Wightman (1992), and Begault (1992b) have shown that there is little perceptual difference between using measured and synthetic HRTFs, where the synthetic versions represent some, but not all, of the spectral details of the original magnitude transfer function.

Asano, Suzuki, and Sone (1990) used a variable pole-zero filter representation of HRTFs, derived from an **auto-regressive moving-average** (**ARMA**) model (a process that results in a data-reduced, synthetic HRTF). Their results from two subjects showed that headphone localization accuracy diminished only with the crudest representation of the HRTF's magnitude. Data-reduced, pole-zero approximations of HRTFs were also explored early on in the work of Gary Kendall and his colleagues at the Northwestern University Computer Music facility (Kendall and Rodgers, 1982; Kendall, Martens, and Wilde, 1990). This technique was found to be acceptable for designing a single filter but unwieldy for designing an entire filter set.

Figures 4.32–4.33 compare impulse responses for three different types of filters. In Figure 4.32, a measured impulse response along with a minimum-phase version of it is shown; note that the peak of the minimum phase impulse response is closer to $n=0$ in time. Figure 4.33 shows the symmetrical impulse response of a linear phase filter. All three of these impulse responses exhibit

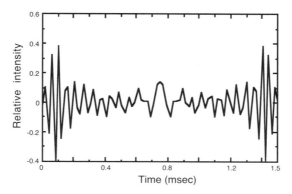

FIGURE 4.33. The symmetrical impulse response of a linear phase, FIR filter.

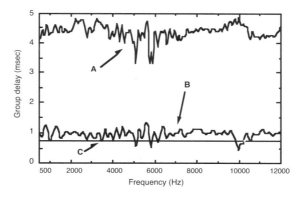

FIGURE 4.34. Group delay plots for measured impulse response A, minimum phase B, and linear phase C FIR filters.

very similar transfer functions for magnitude, but have different phase responses. Figure 4.34 shows the phase response of these filters in terms of their group delay. The raw impulse has a frequency-dependent time delay; the minimum phase versions has a time delay that varies less as a function of frequency; and the linear phase version has a constant time delay across all frequencies. Both psychoacoustic and anecdotal evidence suggests that these filters would be perceptually equivalent for nonindividualized listeners, assuming similar magnitude transfer functions and overall group delay at lower frequencies. This is probably due to the fact the mismatch between measured and synthetic HRTFs is overwhelmed by the localization error inherent to nonindividualized spatial hearing.

Example Procedure for Formulating Synthetic HRTFs

The following describes a procedure used by the author for adapting synthetic HRTFs to a Motorola 56001 DSP chip performing two channels of filtering at a clock speed of 27 MHz and a sampling rate of 50 kHz. A goal was set to design synthetic HRTF filters using 65 FIR filter coefficients for each channel (Begault, 1992b). The steps involved in the procedure were as follows:

1. 50 kHz sample rate HRTF measurements were obtained with a valid frequency response from 200 Hz to 14 kHz. The impulse response of these 50 kHz sample rate measurements normally would be represented by a 512 tap filter.
2. The spectral content of each HRTF was evaluated every 24 Hz using an FFT algorithm, and then stored in a table.
3. For each frequency and magnitude value, a variable weight was assigned, according to a scheme where larger weights were used from 0–14 kHz (i.e., up to the highest usable frequency in the original measurement) and the minimum weight was applied within the frequency range 14 kHz–25 kHz.
4. An FIR filter design algorithm was used to iteratively determine 65 coefficients that best matched the supplied transfer function as a function of the assigned weight (McClellan, Parks, and Rabiner, 1979; Moore, 1986). This algorithm works by distributing the designed filter's error (deviation from specification) across frequency according to the weighting scheme. Since the highest sampled frequency was 11 kHz higher than that of the measured transfer function (14 kHz), the filter was designed so that the maximum error was applied to frequencies between 14–25 kHz in order to better match the measured HRTF below 14 kHz.
5. A constant delay was inserted into the filter used for the ear opposite the sound source, based on the average interaural group delay difference between 200 Hz–6 kHz. Because a linear phase filter design was used, there were no deviations in group delay as a function of frequency.
6. Rather than creating synthetic HRTFs for both the left and right ears, only left-ear HRTFs were used; right-ear HRTFs used the mirror image of the left-ear HRTF. This approach assumes symmetry of the head but reduces the number of synthesized HRTFs needed by half.

Figure 4.35 contrasts one of the measured HRTFs (512 coefficients) with its derived synthetic version (65 coefficients). Note that some of the spectral detail, in terms of the smaller peaks and notches, is lost. A set of 14 synthesized HRTFs produced in this manner deviated no more than ± 5 dB between 400–5500 Hz.

A headphone localization experiment was conducted with thirteen subjects to evaluate the perceptual similarity of the measured and synthetic HRTFs applied to speech stimuli (Begault, 1992b). A within-subjects design was used,

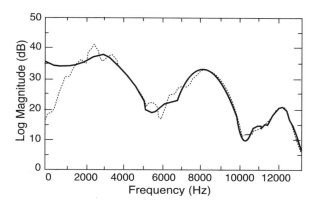

FIGURE 4.35. A measured HRTF (dotted line) and a synthetic HRTF derived from its magnitude Fourier transform (solid line).

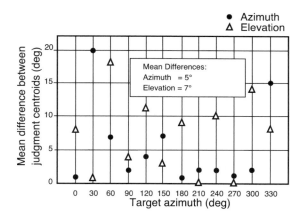

FIGURE 4.36. Difference in degrees for centroids (mean localization judgment vectors), in a within-subject comparison of localization performance with the measured (512 tap) and synthetic (65 tap) HRTFs seen in Figure 4.35. Speech stimuli; means for thirteen subjects (Begault, 1992b).

meaning that each subject was exposed to both types of stimuli. The HRTFs were nonindividualized, representing none of the pinnae of the subjects. Figure 4.36 shows the difference in localization accuracy to be small, 5 to 7 degrees on the average. No statistically significant differences were found between any particular subject's responses to the two types of HRTFs, nor between positions that were obtained by reversing the output channels.

Interpolation, Moving Sources, and Head Movement

Implementation

One of the design considerations for any 3-D sound system is to decide whether or not the simulation interface is constrained to a limited set of measured HRTF positions. The alternative is to allow arbitrary specification of azimuth and elevation via continuous controller or software interface. Additionally, any system that includes head or sound source movement necessitates finer resolution of spatial position than represented by the original measurements. Sound sources typically are not spatially static; there is usually either movement of the head, the sound source, or both. The main exception to this is when headphone playback is used, since in this case a sound source remains static in spite of any head movement. In fact, the absence of body movement interactivity prevents 3-D sound from being as immersive a virtual world as it truly has the potential to be. In Chapter 2 it was pointed out that head movement can allow a listener to improve localization ability on the basis of the comparison of interaural cues over time. Whether or not **source movement** will improve or degrade localization ability is dependent on the movement velocity, type of sound source, and location of movement.

The panpot (panoramic potentiometer) has already been discussed as a means of moving virtual sound sources in space with a traditional mixing console. The amplitude is smoothly weighted from one channel to another, or special functions can be used to create a moving virtual source on the basis of ILD cues. But because a 3-D sound system uses HRTFs, a frequency-dependent change in level must occur when a sound source (or the receiver) moves. Furthermore, changes in the ITD cues contained within the binaural HRTF are necessary.

Because the filter coefficients require intermediate values in order to transition from one stored HRTF measurement to another, the usual procedure is to implement **linear interpolation** between measured values. This procedure is bound to the assumption that an "in-between" HRTF would have in-between spectral features, in the same way that in-between intensities can be obtained by a panpot. Investigation of the moving spectral notches as a function of azimuth or elevation position lends some weight to the premise that spectral features between measured positions can be averaged via interpolation (see Figures 2.19–2.20). However, the result of an interpolation procedure is not always satisfactory, as will be seen subsequently.

If one of the perceptual requirements is to have virtual sound images remain fixed in virtual space independent of where one's head moves, a six degree-of-freedom **tracker system** is necessary, consisting of a **source, sensor,** and a hardware unit housing power supply, electronics, and RS-232 interface to send positional data to a host computer. Currently, tracker technology usually involves electromagnetic sensors, as exemplified by the Polhemus Fastrak and Isotrak or the Ascension Technologies' A Flock of Birds (see Meyer and

Applewhite, 1992, for a review). Alternate systems include optical, mechanical, and acoustic tracking systems, each with advantages and disadvantages in terms of the volume of space one can move in, accuracy, and response latency; most work in 3-D sound has been with the magnetic systems, due to cost, hardware configuration, and the ability to move 360 degrees.

The magnetic tracker's **sensor** is usually attached to the body; those relevant to 3-D sound are placed as close to the top of the head as possible. The sensor measures the low-frequency magnetic field generated by the **source,** which is located outside the body a few feet away and can be used to symbolize the position of the virtual sound source. Positional data about the user's position in relationship to the source can then be obtained in terms of pitch, yaw, and roll, and $x, y,$ and z (see Figure 4.37). The 3-D audio subsystem then can compensate for the relative position of source and sensor to give the illusion of a sound that remains fixed in space independent of head movement.

Figure 4.38 shows the method used by a commercially available 3-D sound system for real-time placement of a virtual source to a fixed position, independent of the listener's movement. The head tracking system sends positional information to the host computer dedicated to the 3-D sound subsystem. An interpolation algorithm constantly checks this position against the two nearest azimuth and elevation positions in memory. The HRTF used at any particular moment is then the interpolation between these prestored values.

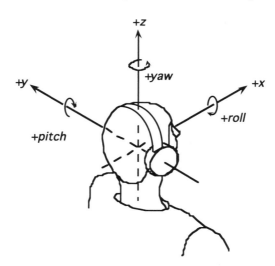

FIGURE 4.37. Coordinates for pitch-yaw-roll and x, y, z in a head-tracked system. *Courtesy of Crystal River Engineering.*

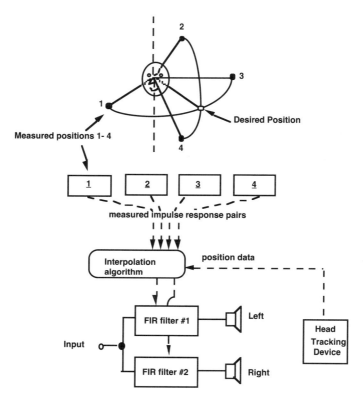

FIGURE 4.38. Interpolation procedure used to compensate for head movement of the listener. To simulate moving sound sources, the position data is obtained algorithmically instead of from a head tracker.

Problems with Interpolation Algorithms

One might at first assume that if one interpolates between two filter impulse responses, one would obtain a valid in-between filter response. For instance, if a filter has a 6 dB dip at 5 kHz for 30 degrees azimuth and a 2 dB dip at 7 kHz at 60 degrees azimuth, the filter at 45 degrees azimuth should have a 4 dB dip at 6 kHz. As Moore (1990) points out, the problem with any time-varying filter is that interpolation is desired in the frequency domain, but in practice the requirement is to interpolate in the time domain. But interpolation in the time domain does not actually yield an average response in the frequency domain. Figures 4.39–4.40 show some examples of how linear interpolation of time domain FIR filters give unexpected results. Figure 4.39 shows the result of linear interpolation in the frequency domain, and Figure 4.40 shows the result in the time domain.

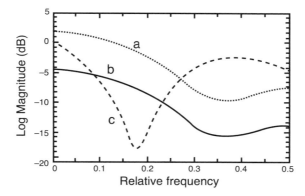

FIGURE 4.39. Linear interpolation of two simple FIR filters, a and b. Ideally, one would obtain an intermediate curve from interpolation that would be inbetween a and b; instead, the curve labeled c results.

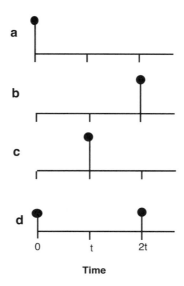

FIGURE 4.40. Consequence of interpolation of time domain impulse responses addressed by Wightman, Kistler, and Arruda (1992). Consider that a and b represent the peak of two measured impulse responses, at 60 and 90 degrees azimuth; the difference in time of arrival consists of the interval 2t. c represents the desired average interpolation value for a 75-degree position, with delay = t. But linear interpolation of A and B result instead in the situation seen in d, two half-height impulses at time 0 and 2t.

The more that acoustic space is sampled and then stored as measured HRTFs within a 3-D sound system, the less significant the error from interpolated HRTF becomes. Most importantly, from a psychoacoustic standpoint, linear interpolation of HRTF magnitude responses seems to be transparent to actual measurement with a high degree of tolerance. A psychoacoustic study by Wenzel and Foster (1993) using minimum-phase filters found that subjects using nonindividualized HRTFs cannot distinguish between signals interpolated between measured sources as far apart as 60 degrees azimuth from sources synthesized using actual measurements. In other words, the error inherent an interpolated HRTF is swamped by the overall localization errors of the subject.

Despite this, synthetic HRTFs designed with minimum-phase or linear-phase filters solve the problem of time domain interpolation (Begault, 1987; Wenzel and Foster, 1993). If one disposes with the need to represent frequency-dependent delays, the ITD portion of the HRTF can be represented with a single value. Interpolation between two single values then will give a correct, averaged delay. The synthetic HRTF techniques used for creating minimum-phase or linear-phase filters from measured HRTFs result in solutions where the group delay is either minimal or zero across all frequencies; the interaural delay information needs to be artificially inserted. The solution is to design the DSP so that one portion of the system implements an interchannel delay equivalent to the ITD of the HRTF and the other portion does HRTF magnitude shaping (see Figure 4.41). These values can be obtained from averaged group delay data (see Figure 4.42).

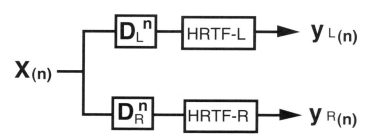

FIGURE 4.41. A DSP system using separate delay and magnitude processing. DLn and DRn are delays used for a particular ITD value. HRTF-L and HRTF-R are minimum or linear phase approximations of measured HRTFs that impart a frequency- dependent amplitude difference on the resulting signal for a given position.

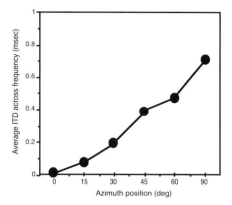

FIGURE 4.42. Averaged ITD across frequency (0.1–6 kHz) between left and right ear HRTFs for one person; 0–90 degree azimuth, 0 degrees elevation (solid dots). Lines show linear interpolation between each value. These delay values can be applied to the D portion of the system shown in Figure 4.41 by converting time into a number of samples. Note that these values are not necessarily ideal for all persons, especially those whose interaural axis is significantly larger or smaller than that of the measured person.

IMPLEMENTING DISTANCE AND REVERBERATION MODELS

Distance Simulation

For distance, an intensity scaling scheme usually is adopted where sounds are emitted at distances *relative* to "egocentric center," inside the head at the center point between the ears. Absolute perceived distances on the other hand are more difficult to predict in an application when the sound pressure level at the ears is an unknown quantity—digital storage and control of playback level is the only option available (e.g., the HEAD acoustics HPS III headphone playback system reads digitally encoded levels to insure constant SPL levels). The discussion in Chapter 3 on IHL points out that an *absence* of interaural attributes places a sound at egocentric position, viz., by sending the same signal to both left and right headphones inputs. Egocentric center can be reserved as a special minimum distance position in its use in an auditory display, for instance as a "sacred space" for only the most important messages (Krueger, 1991, on an idea of Bo Gehring and Tom Furness).

Sound source intensity and its relationship to loudness are discussed in Chapter 3 as basic distance cues. A single digital gain value applied to each output channel from a binaural HRTF filter pair can be used to create a simple "virtual audio distance" controller, according to the inverse square law and perceptual scales shown in Figures 3.2–3.3.

Figure 4.43 shows a graph of dB values that can be used for either perceptual or inverse square law scales for *relative* distance increments. The desired distance increment can be calibrated to a relative dB scale such that it is either equivalent to the inverse square law:

$$20 \log_{10} \left[\frac{1}{\text{distance increment}} \right]$$

or to a perceptual scale approximating the reduction in sones:

$$30 \log_{10} \left[\frac{1}{\text{distance increment}} \right].$$

As an example specification, a distance increment of 0 would be reserved for the loudest, two-channel monaural signal at egocentric center. An increment of 1 would indicate the minimum absolute distance from egocentric center caused by the use of the distance increment scheme, which can be assumed to be somewhere around 4 inches, i.e., in the vicinity of the edge of the head. A distance increment of 2 would double this distance; and so on out to the auditory horizon. These are not the only two distance increment schemes that could be used. In the ultimate 3-D audio system, different perceptual scales could be used according to the type of sound source input and adjusted to each individual's preference. The dB SPL level used for playback could also change the preferred dB scale for a distance increment and the number of increments available before reaching the auditory horizon.

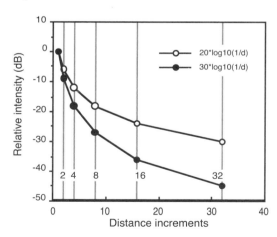

FIGURE 4.43. Intensity increment schemes for relative distance; the inverse square law (open circles) and a $30\log_{10}$ scale that approximates the sone scale.

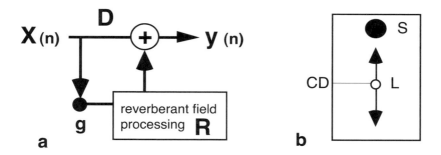

FIGURE 4.44. a: A simple system for implementing the R/D ratio, equivalent to an effect send-receive bus on a standard audio mixer. b: The perceived result of adjusting the R/D ratio is of the listener L moving closer or farther from a sound source S across the critical distance point (CD).

Environmental Context Simulation

Another important cue to distance outlined in Chapter 3 was the R/D ratio, the relative ratio of reverberant to direct sound. The R/D ratio can be implemented by a simple mixing procedure, as shown in Figure 4.44a; the reverberant field can originate either from integrated signal processing or in a distributed system from an off-the-shelf reverberator. Note that this model does not resolve how the R/D ratio, intensity, and environmental context cues *interact* toward the formation of a single judgment for a virtual sound source's distance. Gross manipulation of the R/D ratio will result in a percept similar to Figure 4.44b. It will be seen that the relationship between distance and environmental context cues, based on intensity and reverberation, can all be addressed within a single paradigm. This paradigm involves the techniques of **room modeling** and **auralization**, discussed subsequently.

The implementation of reverberant sound within a 3-D sound system is desirable for many reasons besides distance simulation. Fundamentally, a 3-D sound system without reverberation only can output virtual acoustic simulations of sources heard in an anechoic environment. But the reasons for its use go beyond simulation for its own sake. Because natural spatial hearing usually occurs with reverberation, virtual spatial hearing without reverberation can seem unrealistic (although the same result can be manifested by crudely implemented reverberation schemes). For instance, in the practice of recording engineering, one is almost obligated to manipulate natural and/or synthetic reverberation to make musical instruments sound genuine. For headphone listening, reverberation helps externalize sound sources outside the head. It also provides a sense of the environmental context enclosing the listener and sound source and can give the listener an image of the sound source's extent, as discussed in Chapter 3.

Convolution with Measured Room Impulse Responses

As digital signal processing becomes cheaper, the number of methods and the complexity level for implementing synthetic reverberation based on either measurements or models of acoustic spaces both increase. Some of these methods have been worked into 3-D sound systems, where HRTF processing is applied not only to the direct sound but to the indirect sound field as well, resulting in a synthetic **spatial reverberation.**

It is possible to apply the HRTF measurements paradigm illustrated previously in Figure 4.17 to obtain the impulse response of a reverberant space. The room, rather than the pinnae and the head, are measured through the use of a comparatively louder impulse or maximum-length sequence. One can either measure the room response with a single high-quality (to minimize noise and distortion) omnidirectional microphone or with several microphones at different locations. The result in the time domain will look something like the impulse response of the classroom shown previously in Figure 3.11.

Some of the multiple channel "auditorium simulators" intended for home use are based on a scheme where multiple microphone recordings at the intended locations of the loudspeakers to be used for playback (usually two to four extra speakers in addition to the normal stereo pair). The impulse responses can then be used to convolve the music on your favorite compact disc with the character and directionality of reflected energy in a measured room. One can then supposedly switch the virtual acoustic environmental context between a jazz club, auditorium, gothic cathedral, or other user-defined selection on the simulator. The problem is that most music already contains reverberation so one hears the summed result of the convolution of the original signal, the space it was recorded in, and the user-selected space.

There are several challenges involved in obtaining room impulse response measurements. First, the analytic signal must be loud enough to inject sufficient energy into the room so that the resulting room impulse response has a suitable signal-to-noise ratio: ideally 60 dB above the noise floor of the equipment so that the entire reverberant decay can be measured. Second, the enclosure must have an acceptable level of noise intrusion from outside sources; unless the room is a recording studio there always will be potential sources of error caused by automobile rumble, HVAC (heating, ventilation and air conditioning) systems, and the like. Finally, many rooms have a characteristic reverberation that results from the presence of people; for instance an audience can significantly affect the reverberation time of a concert hall. Because room impulse responses need to be taken under as quiet a condition as possible, one obtains a measurement that simulates an empty room, obfuscating the possibility of capturing some spaces in their normal, occupied condition.

To create spatial reverberation over a headphone playback system, one would need to synthesize the spatial position of reflections if the room impulse response were measured with one or more microphones. HRTF filtering could

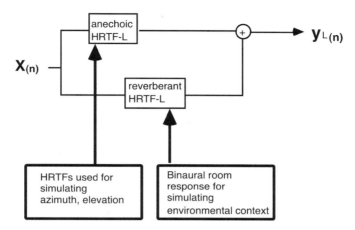

FIGURE 4.45. A hypothetical, yet potentially problematic way of designing a 3-D audio system with spatial reverberation (see text). Only the left channel is shown for clarity. Anechoic HRTFs are taken from a library of measurements, and are used to position the virtual sound source's azimuth and elevation position. The binaural room response originates from a dummy head recording of a room's impulse response (with direct sound removed), and is used to simulate the indirect sound field. Problems may occur if (1) two different HRTF sets are used for the direct and indirect sound filtering, and (2) if the positional information imparted by the reverberation is held constant (as often is necessary with long impulse responses) and the direct sound changes position (e.g., due to a head tracker interface or sound source movement algorithm).

be used to simulate the speaker locations of an auditorium simulator by spatializing the input at each measurement microphone relative to the listener; however, the simulation will be inaccurate because of the **cross-talk** (shared information) between microphones. Interesting results can occur with such processing, but they are far from the results obtained by convolving with actual room responses.

The most accurate means for headphone playback is to measure the room impulse response binaurally. A dummy head recording of an impulse response will capture the spatial distribution of reflections as accurately as it does a direct sound; hence, instead of using omnidirectional microphones, one can collect binaural impulse responses of various rooms at different listening locations, and then apply this to the output of FIR filtering for direct sound, as shown in Figure 4.45. This procedure can result in a formidable number of FIR filter coefficients; for instance, a stereo recording of a two second reverberation time at 50,000 kHz requires a filter capable of 200,000 taps. Although expensive, there are a few commercially available hardware devices, such as the Lake FDP1+

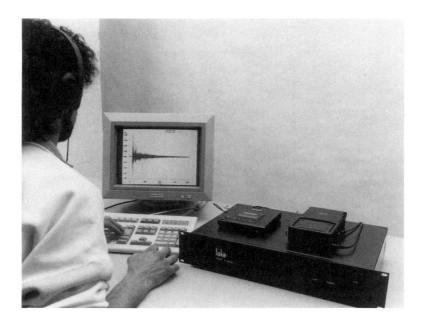

FIGURE 4.46. Specialized hardware for convolution with long FIR sequences include the Lake FDP1+. The user is playing a reflection-free recording through the signal processor and then digitally recording the output while monitoring the result over headphones. The impulse response of the virtual room is seen on the computer screen; this information can originate from either actual room measurements or room modeling programs.

shown in Figure 4.46, that are oriented around long convolution sequences in real time.

Measured binaural impulses of actual reverberant spaces can be modified to create different effects; for instance, a bright-sounding measured room can be low-pass filtered to create a synthetic warm room impulse response, and the reverb time and initial delay gap can be altered via sound editing software. But in addition to the length of the FIR filter needed, other potential problems remain when using an actual binaural impulse response for simulation: (1) it is uncertain whether a match is necessary between the anechoic HRTFs and the pinnae used in gathering the binaural room impulse response (see Figure 4.45); and (2) the position of the direct sound can change, but the reverberant measurement occurs from the same virtual locations unless it is updated too. For these reasons, 3-D sound systems usually are best designed using one of the **synthetic** reverberation schemes discussed later. A synthetic reverberation algorithm can be formed appropriately to send data to either custom DSP systems or via a SCSI interface to special-purpose hardware (e.g., Figure 4.46).

Synthetic Reverberation

Producing synthetic reverberation for years was done by one of three methods: placing a speaker and a microphone in a reverberant room known as an **echo chamber**; sending the signal down one side of a spring or plate and then micing the signal at the other side, the **spring** and **plate reverberator** methods; and finally, the use of **tape head echo** devices. All of these techniques possess a unique sound quality that, while undesirable in representing actual reverberation, are useful to understand and are still used in the professional recording industry to suggest a particular environmental context.

A plate reverberator consists of a rectangular sheet of steel, to which a moving-coil driver and one or two contact mics are attached. Waves transduced through the plate by the driver are reflected back at its edges toward the contact mics; the decay time of the simulated reverberation is a function of the distance of multiple reflections of these waves across the plate. To lower the reverberation time, the plate is partially dampened; to achieve stereo reverberation, two contact microphones are placed at different distances from the driver and plate edges. The spring reverb, found in older organ and guitar amplifiers, used magnetic coils, rather than a driver and a contact mic, and several springs of different length, density, or number of turns. The reverberant effect was created because part of the signal would bounce back and forth within the spring before dissipating into heat, as with the plate. With a reverb chamber, one literally built a highly reverberant room, played the sound to be reverberated with a speaker at one end of the chamber, and picked up the sound with a microphone at the best-sounding location at the other end.

One of the earliest electronic methods for producing a quasireverberant effect was to use tape head feedback with an analog tape system. On a professional analog tape recorder, the erase, record, and playback heads are separate, with the tape passing them in that order. A small time delay results as a function of the tape's movement between the record and playback heads. By returning an attenuated version of the signal received at the playback head to the record head, it is possible to get repeated echo delays that vaguely suggest the delays heard in real reverberation. A reverberant decay could be imitated manually by steadily turning down the volume at g. The system shown in Figure 4.47 is essentially an analog version of the IIR feedback system shown earlier in Figure 4.15, with D being equivalent to the distance between the tape heads. The simple recirculation reverb just described sounds highly artificial (in digital or analog implementation) because of the regularity of the delays and the lack of echo density and because the frequency response consists of a series of alternating peaks and valleys. This "thinned out" frequency magnitude is called a **comb filter** response (see Figure 4.48).

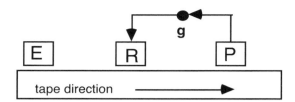

FIGURE 4.47. Tape head reverberation technique. E = erase head; R = record head; P = playback head. This is an analog equivalent of the IIR filter shown in Figure 4.15; the time delay corresponds to the time it takes the tape to travel from R to P. An elaboration of this technique includes multiple playback heads and/or moving playback heads.

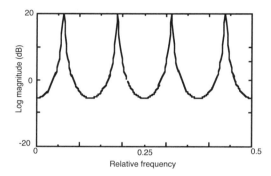

FIGURE 4.48. Comb filter frequency response resulting from a simple delay-feedback technique. The number of resonances corresponds to the delay.

One of the first alternatives to this type of filter was shown by Schroeder and Logan (1961). They set criteria for electronically based "colorless reverberation"; decades later, this term still can apply to reverberators more complicated than the simple IIR feedback loop to which they referred:

> The amplitude response must not exhibit any apparent periodicities. Periodic or comb-like frequency responses produce an unpleasant hollow, reedy, or metallic sound quality and give the impression that the sound is transmitted through a hollow tube or barrel (Schroeder and Logan, p. 193).

Their solution was to use what is termed an **allpass filter** to design their reverberator. The filters that have been introduced in this chapter so far have modified both the magnitude and (in most cases) the phase of an incoming signal. But through an allpass design it is possible for the filter only to affect the phase of the output signal, not its magnitude. In other words, $x(n)$ will

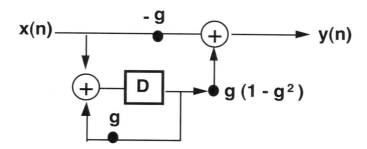

FIGURE 4.49. The allpass reverberator designed by Schroeder and Logan (1961). By controlling the mixture of delayed and undelayed sound and setting D to a particular value, a flat frequency response can be obtained with differential phase.

equal y*(n)* in terms of magnitude—one obtains a flat (linear) frequency response—but the phase response will differ. In effect, the group delay of various harmonics in a complex sound will be skewed in time, in a manner roughly analogous to what happens to the phases of waveforms bouncing back and forth between the surfaces of a room. Figure 4.49 shows the allpass reverberator; Schroeder and Logan (1961) proposed to use several of these units connected in series. As a result, the echo response would become denser and less periodic, i.e., more realistic. This technique is still imitated in many commercially available reverberators. An interesting variant is to use a delay line tapped at different points to feed the input of each of several allpass systems; this has been termed a "nested allpass" reverberator (Gardner, 1992).

Figure 4.50 shows a comb filter with a low-pass filter inserted into the feedback loop. This configuration was used by Moorer (1985) and others to increase density in a frequency-dependent manner; the gains of the low-pass filer could be adjusted to simulate the effects of air absorption. As with the Schroeder-Logan allpass reverberator, several of these units are needed to attain anything close to a realistic reverberation density. Finding the proper values for *D* and *g* can also be quite maddening with systems such as these; trial-and-error, combined with inspection of room impulse responses that are close to the desired virtual room, is about the only way to proceed and can involve years of research.

The reverberator used at a major computer music facility circa 1982 (Stanford's CCRMA—Center for Research in Music and Acoustics) has been described by Sheeline (1982) and Moore (1990). It consisted of algorithms that still can be found in one form or another in commercial digital reverberators. The design represents the culmination of the systems shown in Figures 4.49–4.50. Referring to Figure 4.51, a copy of the direct signal is scaled at *g*, and then fed

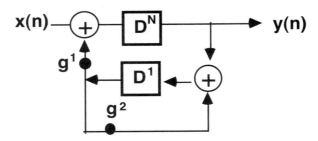

FIGURE 4.50. Comb filter with IIR low-pass filter (adapted from Moore, 1990). Moorer (1985) describes the use of six of these filters, with D^N set to different sample delays equivalent to 50–78 milliseconds. The D^1 and g2 section sets the low-pass frequency cutoff; as g2 is increased from 0 to <1, high frequencies are increasingly attenuated. Higher values of D^N increase the number of resonances in the comb filter spectrum. The intent is to increase the density of reverberation and the complexity of the frequency response.

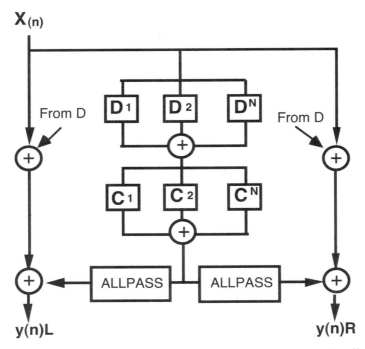

FIGURE 4.51. The Stanford CCRMA reverberator design. C^1–C^N are the comb filters shown in Figure 4.50.

to the reverberation network. A delay line is used to simulate early reflections as a single delay; each value of D could be set to a particular time delay based on a model of early reflections (note that it would be easy to substitute a filter as shown in Figure 4.19 to place 3-D spatialization on each early reflection and/or to represent the absorption of various wall materials). These tapped delays are fed directly to the output. The rest of the reverberator's design focuses on attaining a dense stereo reverberation in an inexpensive manner. The elements labeled $C1–C4$ are low-pass comb filters shown in Figure 4.50; their output is assigned to two or more separate allpass filters. A stereo effect is attained by setting delay and gain parameters for each channel slightly differently.

Convolving an input with noise that has been shaped with an exponentially-decaying amplitude envelope gives a very real reverberation effect if the noise decay matches that of an actual late reverberation response. This method is as computationally expensive as convolving with a real room impulse response, since an FIR filter with thousands of taps is again required. But the technique has excellent sonic characteristics, since the reverberation is dense, and the "comb filtering" characteristics of recirculation-IIR filter techniques are avoided. The approach is also interesting in that it allows different algorithmic approaches to shaping the noise. As pointed out in Chapter 3, reverberation in real rooms does not follow a perfect exponential decay, and the spectral content of the reverberation will vary between different rooms. One can produce digital sound files of noise and then shape them for different effects. Using a desktop computer sound editing program, one can differentially manipulate the rate of decay. One technique useful for creating a stereo effect is to decorrelate the stereo outputs by using differently-shaped amplitude envelope modulation rates for the noise sound files (see Begault, 1992a). The noise can also be filtered to have a particular spectrum in advance of its convolution with a particular sound; for instance, low-pass filtering will make the reverberation warmer and more realistic. Noise other than the white variety can be used, such as pink noise or recordings of wind or water.

Auralization: HRTFs and Early Reflections

Overview of Auralization

A simple working description of a sound source and listener in an environment from Chapter 3 is that sound reaches a listener by a direct path, termed the direct sound, followed by early and late reflections, collectively termed reverberation. In natural spatial hearing, both direct and reverberant sound is altered by the listener's HRTF. In a virtual acoustic simulation, when HRTF filtering is applied not only to the direct sound but also to the indirect sound, the result is **spatial reverberation**. The binaural room impulse response can capture this information, but only within a preexisting room at a predetermined measurement point and at great expense during the measurement process. It is more desirable to obtain a synthetic binaural impulse response characteristic of a real or simulated environmental context so that this information can be used a 3-D sound system for real-time simulation.

Auralization involves the combination of room modeling programs and 3-D sound-processing methods to simulate the reverberant characteristics of a real or modeled room acoustically (see Figure 4.52). "Auralization is the process of rendering audible, by physical or mathematical modeling, the sound field of a source in a space, in such a way as to simulate the binaural listening experience at a given position in a modeled space" (Kleiner, Dalenbäck, and Svensson, 1993). Auralization software/hardware packages will significantly advance the use of acoustical computer-aided design (CAD) programs and, in particular, sound system design software packages. Ideally, acoustical consultants or their clients will be able to listen to the effect of a modification to a room or sound system design and compare different solutions virtually.

There are other uses for auralization that currently are receiving less attention than acoustic CAD but are nonetheless important. This list includes advanced reverberation for professional audio applications (Kendall and Martens, 1984); interactive virtual acoustic environments, based on information from a virtual world seen through a visual display (Foster, Wenzel, and Taylor, 1991); and improvement of localization for virtual acoustic simulations (Begault, 1992a). One can even select a seat in a concert hall by "previewing" its acoustics (Korenaga and Ando, 1993).

The potential acceptance of auralization in the field of acoustical consulting is great because it represents a form of (acoustic) virtual reality that can be attained relatively inexpensively, compared to visual virtual worlds. The power and speed of real-time filtering hardware for auralization simulation will progressively improve while prices decrease because of ongoing development of improved DSPs. Compared to blueprint tracings, an acoustical CAD system allows certain significant physical (and sometimes psychoacoustical) parameters of a speaker-listener-environmental context model to be calculated more efficiently. But it is not necessarily more accurate than noncomputer methods, since the same theoretical base is used for calculation. The results of an auralization system or may not produce anything that sounds like the actual room being modeled because of the difficulty of merging accurate acoustical and perceptual models. The psychoacoustics of many of the issues surrounding room modeling and auralization—in particular the compromises necessary for realizable systems—are just beginning to be evaluated (Jacob, 1989; Begault, 1992c; Jørgensen, Ickler and Jacob, 1993).

Prior to the ubiquitousness of the desktop computer, analysis of acoustical spaces mostly involved architectural blueprint drawings, and/or scale models of the acoustic space to be built. With blueprints, one could draw lines between speakers and listener locations to determine early reflection temporal patterns

FIGURE 4.52. A listener in an anechoic chamber experiencing a virtual concert hall. "Auralization is the process of rendering audible, by physical or mathematical modeling, the sound field of a source in a space, in such a way as to simulate the binaural listening experience at a given position in a modeled space." *Photograph courtesy of Mendel Kleiner.*

and then use a calculator to convert linear distance into time. This type of temporal information is particularly important for speech intelligibility, since loudspeakers in large rooms can be distant enough from listeners to cause delays that exceed temporal fusion (see Figure 2.5). Reverberation time can be calculated from an estimate of the volume of the space and evaluation of absorptive features such as wall surfaces, people and furniture.

One begins in an acoustical CAD system by representing a particular space through a room model. This process is a type of transliteration, from an architect's blueprint to a computer graphic. A series of planar regions must be entered to the software that represent walls, doors, ceilings, and other features of the environmental context. By selecting from a menu, it is usually possible to specify one of several common architectural materials for each plane, such as plaster, wood, or acoustical tiling. Unfortunately, the materials usually are represented very simply, at best usually in octave bands from 0.1–4 kHz. Calculation of the frequency-dependent magnitude and phase transfer function of a surface made of a given material will vary according to the size of the surface and the angle of incidence of the waveform (see Ando, 1985; D'Antonio, Konnert, and Kovitz, 1993).

Once the software has been used to specify the details of a modeled room, sound sources may be placed in the model. Most sound system design programs are marketed by loudspeaker manufacturers, whose particular implementations are oriented around sound reinforcement; therefore, the level of detail available for modeling speakers, including aiming and dispersion information, is relatively high. After the room and speaker parameters have been joined within a modeled environment, details about the listeners can then be indicated, such as their number and position.

Prepared with a completed source-environmental context-listener model, a **synthetic room impulse response** can be obtained from the acoustical CAD program. A specific timing, amplitude, and spatial location for the direct sound and early reflections is obtained, based on the **ray tracing** or **image model** techniques described ahead. Usually, the early reflection response is calculated only up to around 100 msec. The calculation of the late reverberation field usually requires some form of approximation due to computational complexity. A room modeling program becomes an auralization program when the room impulse response is spatialized by a convolution system such as the Lake FDP1+ shown in Figure 4.46.

Figure 4.53 shows the basic process involved in a **computer-based auralization system** intended for headphone audition. Other auralization methods, involving loudspeaker systems or scale modeling techniques, are not discussed here (ref. Kleiner, Dalenbäck, and Svensson, 1993, for a review). A **room modeling program** is used to calculate the **binaural room impulse response**, based on the positions of the source and the listener and details of the environmental context. Note that there is no need to specify

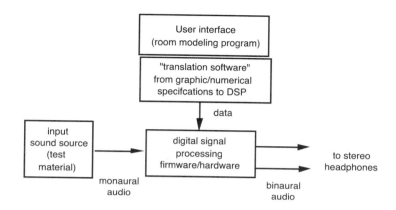

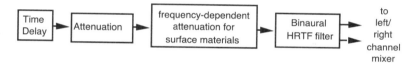

FIGURE 4.53. The basic components of a signal processing system used in an auralization system.

parameters for azimuth, elevation, or distance cues (such as the R/D ratio) explicitly; this data will result as a function of the specified model. The model also will include details about relative orientation and dispersion characteristics of the sound source, information on transfer functions of the room's surfaces, and data on the listener's location, orientation, and HRTFs.

At the bottom of Figure 4.53 is an example of the calculations used in determining the scaling, time delay, and filtering of the direct sound and each modeled early reflection. For each reflection, a "copy" of the direct sound is obtained from a tapped delay line. The path of the reflection from source to listener results in a time delay that is a function of the speed of sound and an attenuation coefficient determined from the inverse square law. The frequency-dependent absorption of the room's surfaces will depend on a complex transfer function that can be approximated by a filter. Finally, the angle of reflection relative to the listener's orientation is simulated by a HRTF filter pair.

Theory

The theory and mathematics involved in modeling sound diffusion and reflection can be extremely complex; only a basic overview will be given here. One can obtain details on mathematical modeling in several sources that are

FIGURE 4.54. Example of sound diffusion from a wall, from a sound source S for a given frequency with a wavelength larger than the wall's surface. The waveform will not bounce off the wall like a beam of light but instead will spread away from the wall in all directions.

relevant to auralization (Schroeder, 1970; Allen and Berkley, 1979; Borish, 1984; Ando, 1985; Lehnert and Blauert, 1992; Kuttruff, 1993; Heinz, 1993).

The modeling of a sound wave as a ray has been used for a long time to describe the behavior of sound in an enclosure and the manner in which it reflects off a surface. This type of modeling is termed **geometrical acoustics.** The analogy is to light rays; one can put aluminum foil on the walls of a room, point a flashlight at a surface, and observe how the beam of light reflects off surrounding surfaces according to a pattern where the angle of incidence equals the angle of reflection. Sound behaves more or less like the light beam to the degree that its energy is **specular**, i.e., the wavefront is smaller than the surface it reflects from. Sound waves that are larger than the reflecting surfaces are by contrast **diffused** outward in a complex manner (see Figure 4.54). Consequently, many sounds, which are complex and contain both high and low frequencies, will be partly diffused and partly reflected.

Figure 4.55 shows the **image model** method for calculating early reflection patterns in two dimensions (Allen and Berkley, 1979; Borish, 1984). In practice, a three-dimensional model is used to determine patterns from walls, ceiling, and the floor. The idea is to model sound as a single specular reflection to each surface of the modeled room, as if the sound were a light beam within a "room of mirrors." By finding the virtual sound source position in a mirror image room, a vector can be drawn from the virtual source to the receiver that crosses the reflection point. Perhaps the first implementation of the image model technique with HRTF filtering was the "spatial reverberator" developed by Kendall and Martens (1984) (see Figure 4.56).

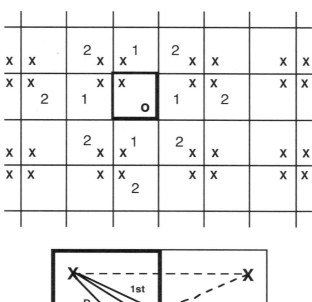

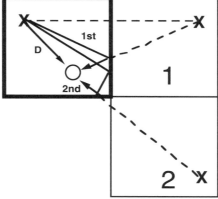

FIGURE 4.55. Use of the image model method to calculate early reflection patterns (Allen and Berkley, 1979; Borish, 1984; Kendall, Martens, and Wilde, 1990). X = location of actual and virtual sound sources; O = location of the listener. Top: A slice through the three-dimensional image space; the source in the actual room (shown with the bold lines) is mirrored outward to create first-order reflections (labeled **1**); mirrors of mirrors create second-order reflections (labeled **2**), etc.. Bottom: magnification of upper illustration, showing direct sound (D) along with a first-order reflection (one bounce) and a second-order reflection (two bounces). Dashed lines through the real and mirrored images show the reflection point on the wall.

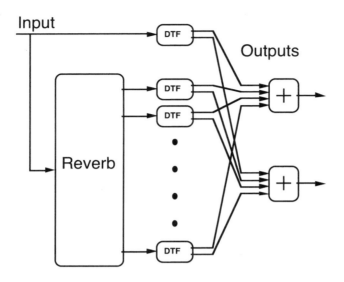

FIGURE 4.56. The spatial reverberator developed at the Northwestern University Computer Music facility in the 1980s (Kendall, Martens and Wilde, 1990). The reverberation is calculated according to an image model and then direct and reverberant sounds are sent through a bank of HRTF filters (here, DTFs—"directional transfer function" filters). *Courtesy William L. Martens.*

The image model can be extended to successive orders of reflections; successive bounces are calculated by extending outwards with "mirrors of mirrors," etc.. Typically, auralization systems use no more than second or third order reflections, since the number increases geometrically with each successive order (see Chapter 3). The process works well for simple room models, but for complex shapes akin to real rooms reentrant angles will be obstructed by surfaces. Rays may or may not encounter a listener placed in the model, requiring adjustment of any final calculations (see Borish, 1984; Lehnert and Blauert, 1992).

Another method for determining early reflections is to use a **ray tracing** process (see Figure 4.57). In this method, the sound source is modeled as emitting sound "particles" in all directions; the particles can be approximated by rays emitted from a sound source, according to its pattern of emission (Schroeder, 1970). But unlike the image model method, where a first-order reflection is calculated once to each surface, the ray tracing method calculates an energy distribution that is applied many times to each surface. As each ray encounters a surface, it can be appropriately filtered according to angle of incidence and wall absorption.

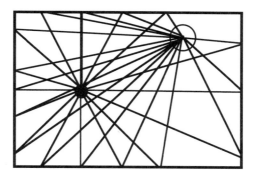

FIGURE 4.57. A simple two-dimensional ray tracing model.

Ray tracing has been used for some time to obtain accurate measurements of reverberation time. Unlike the image model, the effects of diffusion can be calculated more accurately with the ray tracing method, but it is much more computationally intensive to perform because a large number of rays usually are needed for accurate results. Some auralization systems use combined image model–ray tracing schemes. Other computationally intensive but more accurate techniques, such as **boundary element** or **finite element** modeling, may find increased use in the future as computation speed increases for the desktop computer.

Implementation

Figure 4.58 shows a heuristic example of auralization signal processing as applied to two spatialized early reflections. Consider a sound source at 0 degrees; a reflection incident from left 60 degrees delayed by 10 milliseconds and 4.5 dB down relative to the direct sound; and a second reflection incident from right 60 degrees, delayed by 14 milliseconds and 6 dB down. One could simulate this sound source by combining the binaural impulse responses for each channel shown in Figure 4.58, and then using the result in an FIR filter structure such as that shown in Figure 4.19. These figures represent the basic idea of the signal processing involved in forming spatial reverberation, except that a larger number of impulse responses are combined to represent additional reflections. Often, due to the computational expense of FIR filtering, the process is applied only to early reflections and a synthetic reverberation process such as those discussed earlier is used for late reflections.

The reverberation simulation used by Begault (1992a) consisted of two parts: an early reflection pattern and a late reverberation pattern, based on a model of a listener one meter distant from an omnidirectional sound source. The direct sound was also HRTF filtered in the implementation. The early reflection pattern was calculated using a simple two-dimensional ray tracing model

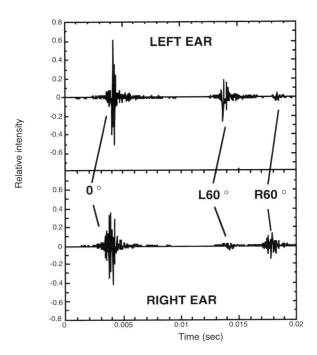

FIGURE 4.58. Summation of HRTF impulse responses to form spatialized reverberation using a direct sound and two reflections. The entire response for each ear could be used in an FIR filter such as shown in Figure 4.19, with 1,000 taps per channel (50 kHz sample rate).

similar to that shown in Figure 4.57. The late reverberation portion of the stimuli was modeled as two separate distributions of exponentially decaying noise, one for each output channel. Slight differences were implemented to decrease correlation between the channels for greater realism in the simulation. The noise was generated digitally with a pseudo-random number generator as a mixture of white and 1/f noise (1:1 ratio for the left channel, 3:2 ratio in the right channel).

The signal processing involved sequential steps of FIR filtering of the direct sound according to the model described previously in Figure 4.53. For each of the modeled early reflections, the angle of incidence to the reflecting surface, absorptiveness of the reflecting surface, and path length given by the ray tracing program determined an initial attenuation and time delay of a copy of the direct sound. Subsequently, the angle of incidence to the listener was calculated, and the nearest measured HRTF (every 30 degrees azimuth) was used to create a two-channel, spatialized early reflection. The impulse responses for all early

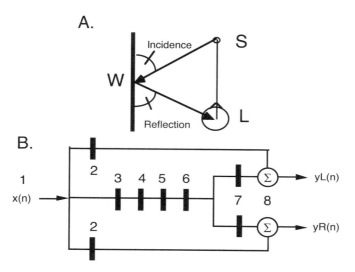

FIGURE 4.59. Calculation of parameters based on a ray tracing algorithm.

reflections were then summed for filtering. Finally, the left and right outputs were convolved with the noise signals for simulating late reverberation.

A detailed explanation of the signal-processing system is shown in Figure 4.59. At the top is a simplified illustration of a single reflection and a direct source, comprised of a sound source S, a wall W, and a listener L. At the bottom is a diagram of the digital signal-processing algorithm; the numbers show each stage of calculation based on the ray tracing program, corresponding to the numbers in the following description.

1. The digital sound source input $x(n)$ is split into three paths. The center path involves the processing of the early reflections, based on the ray tracing method. Each reflection corresponds to the path lines S-W-L in Figure 4.59a. The upper and lower signal processing paths in Figure 4.59b correspond to the direct sound shown as the line S-L in Figure 4.59a.
2. The direct sound is spatialized by a single HRTF filter pair corresponding to the angle of incidence of the direct sound to the listener.
3. An attenuation based on the inverse square law and a delay based on sound velocity (334 m/sec) are calculated for the distance of the sound path S-W.
4. A magnitude transfer function corresponding to the approximate frequency-dependent characteristic of a specified surface is convolved with the signal.
5. The angle of reflection from the surface is analyzed and then the signal is further attenuated according to a model of specular reflection, such that for

frequencies above 500 Hz, amplitude is decreased as the angle of incidence of the reflection from the surface *W* becomes smaller.

6. An attenuation and delay is applied for the path *W-L*, in the same manner as in step 3.

7. The angle of incidence of the reflected sound to the listener's modeled orientation is calculated and then used to determine which HRTF filter pair is used to spatialize the reflection.

8. Each channel of the resulting spatialized reflection is summed with the direct sound.

Summary

The implementation of environmental context simulation in a 3-D audio system requires a method for convolving an input signal with spatialized early reflections and reverberation. The synthetic reverberation method allows an intuitive approach, but it is difficult to match a realistic natural spatial hearing condition. Measured binaural room impulse responses represent only one room convolved through a particular set of HRTFs and are difficult to modify once they are obtained. The auralization approach to environmental context synthesis represents a good compromise between the two other methods, but the simulation will be only be realistic if a sufficient number of accurately-modeled parameters are accounted for.

CHAPTER FIVE

Virtual Acoustic Applications

INTRODUCTION

There are many uses for 3-D sound that are known already in the entertainment, broadcasting, and virtual simulation application areas. Many undiscovered uses probably exist as well that will entail either variations on existing system designs or entirely new approaches. What follows is a look at only a few of these possible application areas, with both integrated and distributed 3-D audio systems. Most of these systems actually exist and are operable; some actually are used as tools and not merely as demonstrations of an interesting concept or foggy fantasies that exist only on paper. Note, however, that development is almost impossible without ideation and fantasy that transmutes into hastily assembled concept demonstrations, prototype systems, and first releases. By watching or studying (in a formal experiment) the way a user interacts with a new implementation of 3-D sound, one discovers not only human factors issues that remain to be solved but also unexpected advantages that could never have been predicted. Although expensive and far from optimal, a prototype system allows room for *discovery* of new applications and ideas and does not just provide a checklist of how to improve the second prototype.

The use of 3-D sound within a particular application requires sound input; and in a distributed system, the MIDI standard is frequently used. For that reason, this chapter begins with these topics and shows their integration into an example application. Following this, specific applications are reviewed, concluding with the aesthetic use of 3-D sound for artistic purposes in computer music.

SPEECH AND NONSPEECH AUDIO INPUT

A review of 3-D sound techniques cannot focus exclusively on the methods that cause sound to be heard spatially, since some form of audio *input* ultimately becomes relevant. Spatial hearing, of course, is very dependent on physical transformations and psychoacoustic cues, but in an application, part of the success will depend on the sound that is transformed. To paraphrase an old saying about computers, what comes out of a 3-D audio system will be only as good as what one puts into it. The type of input to be used should be guided by the requirements of the intended application and the available resources, as covered in Chapter 1 and Chapter 4.

Sound input can be segregated between speech and nonspeech audio (an in-between area, **speech synthesis,** is a topic simply too complex to cover here). The most important criteria for speech transmission is maintaining intelligibility throughout the entire communication system chain. With nonspeech audio, the functions and means for producing sound are more varied. Buxton, Gaver, and Bly (1989) categorize nonspeech audio into three categories: **Alarms and warning messages, status and monitoring messages**, and **encoded messages**.

Alarms and warning messages function not only from automobile horns or system beeps but are important features of complex human interfaces. Alarms become particularly important in high-stress environments, such as airline cockpits. In fact, until the 1980s and the advent of synthesized talking instrument panels, many cockpits used as many as sixteen different nonspeech warning signals that each had a specific meaning. We know from the work of Patterson (1982) and others that the best design for an alarm in such situations is not necessarily a loud, jarring bell or something similarly stress-inducing.

Currently, an international standard is being drafted for the multitude of alarms and other auditory signals that are used in health care equipment, particularly in hospitals. At present the alarms are based on the louder-is-better paradigm, a situation that makes it difficult for people to talk in situations where communication is important. This is a situation where well-designed, easily localizable auditory signals could be put to good use. But several problems have yet to be overcome. First, the machines hooked up to alarms cannot distinguish context: A removed breathing tube will sound an alarm whether it was removed by the doctor or by accident. Second, multiple alarms can sound from different devices in response to a single change in the patient's condition, such as a drop in blood pressure. Particularly telling is the comment that "These machines are all around the room, and with so many of them, when an alarm sounds, often you don't know where it's coming from" (Van, 1994, quoting Dr. Peter Lichtenthal of Northwestern University Medical School).

Status and monitoring messages are described by Buxton, Gaver, and Bly (1989) as the type of ongoing sound that usually is only noticed when a change occurs in the state of a task or process. This type of information is also where simple **auditory feedback** from an action or a process is used. An everyday example is an automobile's engine; its sound gives continual information about its current state and we notice it only when it's time to shift gears or when something's wrong. The act of lighting a gas stove with a match is a more complex example of auditory feedback; the amount of gas being discharged and whether or not the flame is lit, can be distinguished more safely by listening than by other means. Auditory feedback in both aeronautical and virtual reality contexts is covered in detail later in this chapter.

In some cases, modern technology has taken away auditory feedback from the operator and then needed to restore it later. A case in point is the typewriter versus the computer keyboard; those who have worked professionally on the ubiquitous IBM Selectric typewriter of the 1960s and 1970s knew it to be a loud device with several intended and unintended sources of auditory feedback (as well as considerable subaudio vibration that caused everything within twenty feet to become a vibration source). With the advent of personal computers came new, quiet keyboards. However, later designers were forced to add noise back into computer keyboards, since a lack of aural feedback in connection with tactile action was found "unnatural" to some. On some computers, one can even add a special "typewriter click" sound to the normal keyboard sound.

An example of encoded messages, sometimes referred to as data **sonification,** involves the use of variations in continuous sound for data exploration (Bly, 1982; Mezrich, Frysinger, and Slivjanovski, 1984; Smith, Bergeron, and Grinstein, 1990; Scaletti and Craig, 1991). Here, feedback is not used for verifying physical actions but for acoustically monitoring the contents and relationships within a database that are usually shown graphically. **Multivariate data,** measurements where several variables interact simultaneously, are usually conveyed via graphical means, but only a limited number of dimensions can be taken into a single graph. As Bly (1982) points out, when dealing with multivariate data about which very little is known, alternative or supplementary means of representation can be helpful. Since sound has perceptually segregatable attributes of pitch, loudness, timbre, and spatial location, and is organized over time, data where time is a variable might be best heard rather than seen. This can be done in a crude way by identifying which dimension of a tone generator—pitch controls, intensity controls, filters, etc.—is associated with each data variable.

For instance, suppose we work with a public health department and need to understand the interaction of the following variables: time, air temperature, smog content, and number of cars that pass a particular point on the freeway. A rough view of their interaction might be heard more quickly with a 3-D sound system, compared to visually inspecting several x-y graphs where each variable is paired. For example, the smog level can be connected to pitch, the number of cars to loudness, temperature to spatial position left-right, and time might be compressed so that one hour takes thirty seconds. With such an audio-graph and a fair amount of training, it would be possible to hear in just six minutes whether a change in smog level (pitch) was related more to the number of cars (loudness) or temperature (rightward movement) across a twelve hour day. The successful implementation of such systems faces a complex set of challenges, since a single percept can be affected by several physical factors. As reviewed in Chapter 1, two different frequencies of the same objective intensity can be heard at different loudness, and so on.

The range of sounds that can be produced by various sound synthesis techniques is enormous. For the most part, the tools used for nonspeech synthesis come from the computer music and sound effects world. Generally, most of the nonspeech sounds we hear over loudspeakers and headphones come either from **tone generators** (synthesizers that create and modify waveforms from scratch) or **samplers** (where the basic waveform is recorded from a real source and then processed). On many multimedia plug-in cards, for example, one can use **frequency modulation (FM) synthesis** to approximate actual sounds. But the overall lack of flexibility of these systems potentially can make them unsatisfying for application to the real time synthesis and modulation demands of input auditory signals. More complex, albeit more difficult approaches to synthesizing sound include software synthesis and complex MIDI control (e.g., HMSL, Csound, cmusic, MAX™, and others). An impressive recent entry that is quite amenable to sonification applications is the Kyma™ system, available from Symbolic Sound Corporation of Champaign, Illinois; it consists of a hardware system ("Capybara") that uses a Macintosh platform and up to nine separate DSPs and a visually based software language ("Kyma"). Their product literature is worth quoting: "Anticipate the next wave. Data sonification is a new area, ripe for exploration. Data-driven sound will be an essential component of multi-media presentation and virtual reality construction. Now is the time for perceptual psychologists, visualization specialists, and others familiar with electro-acoustic music to begin developing the techniques and the tools that will allow us to use our ears as well as our eyes in exploring and interpreting complex data" (Kyma, 1991).

Still, because the most flexible sound-producing devices available anywhere still can produce sounds that annoy the end user after continued use, several writers have pointed out the necessity of including a sound design expert in any sonification design team. There simply isn't an exact science that allows one to predict the success of any signal used in a virtual acoustic display, and if there were one, the end user would want to change between various configurations at will, as evidenced by the way that some users add and change numerous types of personalized system beeps and other sounds to their computers.

Representational Sounds

The development of an auditory symbology need not lie strictly with sounds developed from simple waveforms; sampled sounds can be collected and used for their purely representative values in acoustic interfaces. The term **auditory icon** was meant to specifically represent Gaver's (1986) harnessing of **everyday sounds** at the computer interface. According to Blattner, Smikawa, and Greenberg (1989), an auditory icon is a type of **earcon**; earcons also can include abstract, synthesized sounds or combinations of the two sounds. The argument for using everyday sounds is that nonmusically trained or otherwise acoustically unsophisticated persons do remarkably well in identifying, categorizing, and classifying the sounds all around us; therefore, the use of everyday sounds in a human interface would accelerate the learning process for making good use of them (Gaver, 1986; Buxton, Gaver, and Bly, 1989). In an example from the Sonic Finder for the Macintosh, Gaver (1989) attached tapping sounds when a folder was opened that varied in pitch as a function of how full the folder was. Dragging a file to the visual trash can icon used by the Macintosh operating system caused the sound of something being dumped in an actual trash can.

The harnessing of the cognitive yet subtle power of sound sources has been long known to sound effect (**Foley sound**) artists in the movie industry. In almost any commercial film, the background sounds (those in addition to dialogue and music) are synthetically formed and inserted at a later point. The use of sound effects becomes an art through the difficult process of matching a particular sound to the emotional setting called for by the cinematography and the script. Sound effect artists amass huge collections of recordings of everyday sounds, such as footsteps, doors opening, machinery, and background sounds. In many cases, recordings of unlikely objects are used; dog food being ejected from a can was used in a one major film release to suggest a surreal physical transformation. Less difficult is the process of making the sound believable, since listeners often have a willing suspension of disbelief with sound that's equivalent to the experience of drama or opera. For example, to make a punching sound for a fist-fight scene, one artist mixed the following sources: a gunshot (for the impact), a sound of rushing wind (for the fist traveling through the air), and a frozen turkey being slapped (to give the overall sound a denser, richer quality). The actual sound of physically damaging contact in a fist fight is rather dull, not very loud; therefore, the sound effects artist must make it exciting.

Our perceptual system is often explained as being dominated by vision— indeed, although the TV speaker may be displaced from the tube or the movie theater speaker from the screen, we imagine the sound as coming from an actor's mouth or from the car. But there are many cases where sound can powerfully influence the perception of visual images. A good example is the commercial for a mail-order briefcase that was featured on late-night television some years ago. The visual sequence consisted of an executive opening and closing a flimsy, phony leather briefcase, maybe forty times in the space of two minutes, with close-ups of fingers operating the buttons that opened the latches. A narrator's voice simultaneously expounded on the fine quality and strength of what was obviously a low-quality product. Behind the narrator's voice, however, was a sound that was far from flimsy. A loud, mighty, robust latching sound—sampled from something akin to a close-mic recording of a closing bank vault door—was activated with each opening and closing of the briefcase. Some viewers probably bought the mail order briefcase because it **sounded** like quality, in spite of the fact that a cheap three-inch speaker was the ultimate transducer.

An Aeronautical Application of Auditory Icons

Applying auditory feedback along with spatial information makes excellent sense in a high-stress human interface such as a commercial airliner cockpit. In a human factors study designed to test the suitability of new control panel interfaces, an aerospace human factors researcher group evaluated commercial airline pilots' use of a touch screen computer panel. This panel displayed both checklists and flight control devices in place of paper checklists and normal switches (Palmer and Degani, 1991). The study took place in an experimental "advanced cab" flight simulator used for evaluating new technologies at NASA Ames Research Center.

A problem found at the outset was that pilots were uncertain when they had positively engaged the "virtual switch" on a touch panel computer screens mounted into the aircraft control panel. There was also a problem in the lack of a perceptible difference between merely "sliding across" several switches to an intended target switch versus actually engaging the switch. Finally, a small but noticeable time lag between tactile action and visual feedback caused by the processing time needed for updating the graphic displays, made interaction with the touch panel frustrating.

A solution by the author (Begault, Stein, and Loesche, 1991) was to link the touch screen's virtual buttons to an stereo audio sampler via MIDI control, so that prespatialized, representational auditory icons could be played though high-quality stereo communication headsets. Several types of sounds were recorded into the sampler that were activated in relation to different switch functions on the touch panel. Both pilots could access the touch panel during flight operations via separate screens. The sounds were spatialized using 3-D audio techniques, so that the copilot's auditory icons came from the right side of both pilots, and the pilot's auditory icons came from both pilot's left side. This arrangement allowed a situational awareness that informed one pilot if the other was engaged in menu selection so that neither needed to turn away from more critical operations to make visual verification. The overall amplitude envelope of the auditory icons was processed so that their levels and spectral content would easily penetrate ambient noise and radio communication speech but would not be annoyingly loud after continual use.

Figure 5.1 shows two different examples of the "soft display" touch panel pages. There are four touch-sensitive panels used for displaying both automated checklists and engine-electrical-environmental controls of the aircraft; functions are selected by touching appropriate virtual switches on one of several menu displays. An auditory distinction was made between three classes of switch functions so that actions related to the operation of the aircraft were distinguished from actions related to selecting items from the menu. The default sounds for menu-selection button operations used two "click" recordings of a high-quality aircraft switch (engaged and disengaged). The switch has a very different sound when pushed in compared to when it is released; the push-in sound was used for making finger contact on the virtual switch, while the release sound was used when the switch was actually engaged.

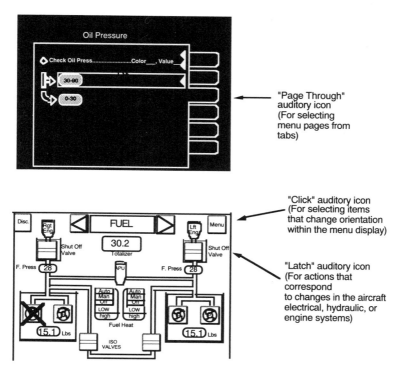

FIGURE 5.1. Touch panel pages from the advanced cab cockpit simulator at NASA Ames Research Center, showing how different auditory icons are associated with specific actions. The "latch" icon is distinguished perceptually from the "page through" and "click" icons by use of a louder, more noticeable sound, because it corresponds to an action that changes the operating state of the aircraft.

A recording of index cards being shuffled initially was used for the menu page through control but later was replaced by the default click sounds. This change was because the duration necessary for meaningful identification of the sound was too long (shuffling takes longer than clicking). This case is a good example of how an auditory icon can be representative of an everyday sound related to more-or-less abstract interpretations of the visual screen metaphor. A noticeably different auditory icon was used for actions corresponding to a change in the flight system. A latch auditory icon, using a recording of a high-quality door lock mechanism, was used for engaging fuel pumps and other activity directly related to the aircraft system. This auditory icon was processed to be louder, longer in duration, and timbrally deeper (via a bass boost) relative to the click and page through icons, in order to provide a perceptual distinctiveness and special significance to the action.

HRTF filtering was used to spatialize recordings of the engaged-disengaged button sounds in advance; separate MIDI note-on instructions were used to spatialize the left panel buttons to the left 90 degrees and the right panel to the right 90 degrees for both pilots. Informal observation seemed to suggest that adding auditory feedback in response to tactile actions increased performance and user satisfaction with the virtual controls. It also seemed as though you could feel the switch being engaged in and out, simply by virtue of the proper sound. This experience is a type of **synthesisia**, a cross-perceptual modality sensory substitution.

A Case Example: Illustrating MIDI Communication

For communicating between distributed hardware devices, the use of the MIDI standard is often convenient. Most currently available tone generators, samplers and computers have MIDI interface capability and an increasing number of mixers and peripheral devices use the standard. There are also RS-232-to-MIDI converters available if one wishes to program directly to the computer port. The MIDI standard is also convenient because multiple devices can be connected in series and then instructions can be relayed along a specific **MIDI channel.** Only devices set to receive on the particular channel sent will be activated. There is also an **omni mode** that allows reception of any channel instruction.

It is necessary to have a good understanding of how to write digital control signal code in order to successfully interface various devices in a distributed 3-D audio system. Fortunately, the MIDI standard is implemented in most devices, and there are specific libraries (usually in the C programming language) one can add to a customized C program routine. The MIDI standard is also relatively simple, since provision for **control signals** are matched to audio synthesizer hardware built for music and audio production, where real-time control of audio parameters is often necessary. Technically, MIDI is a real-time, 31,250 bit/second serial interface, adapted as an international standard for transmitting event messages and a timing reference in a device-independent manner between microprocessor-equipped musical instruments and computers.

Consider a simple example implementation where MIDI is used to communicate instructions to "backend" audio and sound spatialization servers from a computer operating system. We want a spatialized click to sound when the mouse button is engaged. For heuristic purposes, the computer has a MIDI port, but no tone generator or 3-D sound system; these exist as separate hardware devices. The 3-D audio subsystem is specified to match the position of a mouse click anywhere on a computer monitor screen, relative to an arbitrary audio mapping position. Figure 5.2 shows the screen designated into nine equal-sized regions with corresponding auditory positions, three azimuths by three elevations. The software must read the x-y coordinates of the mouse pointer on the screen, determine which of the nine regions the coordinates fall into, and then produce a spatialized signal corresponding to the auditory mapping.

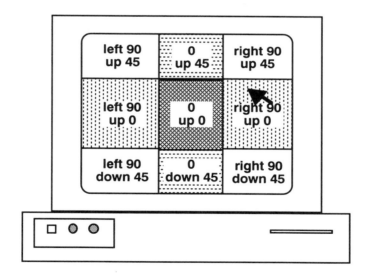

FIGURE 5.2. A simple mapping of mouse position on a computer screen to a spatialized beep.

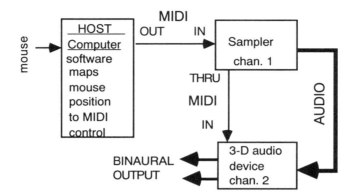

FIGURE 5.3 Illustration of the control signal paths for a mouse–audio interface showing how multiple devices can be addressed by MIDI control signals. Tracing the audio signal paths, a monaural signal is generated by the sampler, and is then spatialized into a two-channel binaural output by the distributed 3-D audio device for output to stereo headphones. Tracing the control signal paths, the computer is used as a MIDI controller and the sampler and 3-D audio devices act as MIDI slaves. The slaves are connected in series, but can be addressed independently due to their unique MIDI channel number. The mouse position is read as control signals to the host computer's operating system; a software program reinterprets this into commands that are sent out the computer's MIDI out port to the MIDI in port of the sampler, which responds to information addressed to MIDI channel 1. The MIDI thru port of the sampler passes instructions from the computer unchanged to the MIDI in port of the 3-D audio device, which is set to respond only to MIDI channel 2.

Figure 5.3 shows the configuration of the hardware devices. A monaural signal is generated by the sampler and is spatialized into a two-channel binaural output by the 3-D audio device (thick lines in Figure 5.3). To activate the external devices from the computer, MIDI instructions are sent along a series-connected set of devices, each with its own unique MIDI channel number. This **control signal** path is to be distinguished from the audio signal path; you can only hear the effect of a control signal on an audio signal. Note that in an integrated 3-D sound system (e.g., communication between the 3-D audio and tone generator chips on a plug-in multi-media sound card), the control signals are sent internally.

In this example, sending a MIDI command to play middle C on the sampler activates the mouse click sound. It is only necessary to send the MIDI equivalent of a **trigger** to the sampler. But whereas an analog trigger is simply an instantaneous transition in voltage level that conveys no other information, the MIDI message is more complex. For this, one uses a note-on command that consists of the three 8-bit words (bytes) shown in Figure 5.4. The first

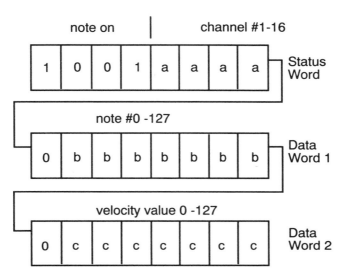

FIGURE 5.4. Anatomy of a MIDI message for triggering a device. The message is grouped into three eight-bit words that are evaluated sequentially: the status word, followed by two data words. The status word 1001 used here indicates a note-on command in the first four bits, and the MIDI channel number (1-16) on the last four bits. The first data word uses the last 7 bits to indicate a note number from 0–127. The second data word indicates a velocity value between 0–127, directly analogous to a control for relative volume.

information in the transmitted data is identification of what kind of message is occurring and on what MIDI channel number; this is represented by the **status word** byte in Figure 5.4. Next, two **data words** are sent: first, the note number (0–127; middle C is equal to 60) and then the velocity value that represents how loudly the note is played (0–127, min–max). Any tone generator that receives this note-on instruction command at its MIDI in port will produce the requested sound, so long as its receive channel matches the channel indicated in the status word.

For sending information on the position of a continuous controller, such as a modulation wheel on a synthesizer, a slider control, or a foot pedal, a different type of MIDI message than a note-on is used. Figure 5.5 shows how a MIDI **control change** message is transmitted. The status byte 1011 identifies the information following as control change data and again indicates a unique MIDI channel number in the last four bytes. The first data word is used for a unique controller number (tone generators can respond to several controllers in real time), and the second data word indicates the value corresponding to the position of the controller (e.g., the slider at maximum position would equal 127).

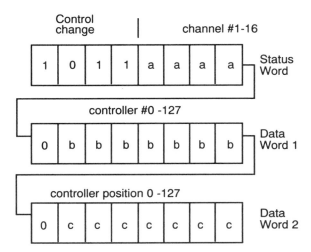

FIGURE 5.5. Anatomy of a MIDI control change message for transmitting data from continuous controllers. In this case, the controller number is mapped to a particular function on the MIDI device (e.g., azimuth control input), while the position is mapped to the value for that particular function (e.g., where on the azimuth).

While it is more typical to use MIDI controller data for manipulating the familiar features of tone generators, such as volume control or modulation depth, there is no reason why more exotic controls cannot be effected in real time, such as reverberation levels, lighting, and MIDI-controlled popcorn poppers. In fact, the use of MIDI controller information is exploited in some off-the-shelf 3-D sound devices that allow one to fly around in azimuth as a function of moving the modulation wheel on a synthesizer, a pedal, or any MIDI output device.

To spatialize the mouse click sound, the 3-D sound system first must be set up to the proper spatial coordinates. Assume for this example that the 3-D sound device receives MIDI in such a way that controller number 100 can be mapped to azimuth and controller number 101 to elevation. It is then necessary to send two separate MIDI control change messages for a given spatialized position. We can then set up something like Table 5.1 as a guide to programming the correct information for data words 1 and 2 in the MIDI control change command corresponding to the spatial regions mapped onto the computer screen in Figure 5.2.

Using the information from Table 5.1, a computer program can be written for spatializing mouse clicks that might resemble the C language code shown in Figure 5.6. An upper-level routine checks the position of the mouse when it is clicked, and returns an x-y value corresponding to its pixel location. The sound management routine then takes this position, tests which of the nine regions of the screen that the mouse clicked on, and sends appropriate set-up

Controller no. (Data Word #1)	Function	Position	Controller value (data word #2)
100	azimuth	left 90°	0
		0°	64
		right 90°	127
101	elevation	down 45°	0
		0°	64
		up 45°	127

TABLE 5.1. Mapping MIDI data to computer screen regions. The actual values used for data words 1 and 2 are arbitrary.

information to the 3-D sound device over MIDI channel 2. The command send_midi_control takes three arguments for sending control change information: the MIDI channel number, the controller number, and the controller position, which are obtained from Table 5.1. After the 3-D sound device is set up, the sound can be activated. The command send_midi_note_on sends the following arguments to the tone generator: MIDI channel number, the note number, and the velocity. In this example, the velocity is maximum, so users should have their headphone volume rather low. The commands send_midi_control and send_midi_note_on would call lower-level commands (often included in a library) that are designed to generate the binary version of the MIDI information at the serial MIDI port.

This review is only a basic description of what MIDI can do; for the interested reader, there are many published overviews that range in level and sophistication, from how to hook up two MIDI devices to how to write custom computer code to fully exploit MIDI's capabilities (see Chapter 6). Note that MIDI is certainly not the only way to interface distributed devices; for instance, at short distances, parallel rather than serial ports are preferred to avoid timing problems that creep up as MIDI instructions become more complex (for an interesting critical paper, see Moore, 1987). Note also that the degree to which one can use numerous MIDI features will be constrained by both the capability of the devices to receive MIDI information and the kind of real-time control desired for transmission.

```
// UPPER LEVEL ROUTINE; checks to see if
// mouse is down and, if true, calls subroutine
// to obtain position on 640x480 pixel screen
// (values of x and y, with 0,0 at upper left corner.

  main()
  {
   int x, y;
   if mouse_click( ){
       get_mouse_position(andx,andy);
       play_sound(x,y);
  }

  // SOUND MANAGEMENT ROUTINE; take mouse x,y
  // values, set up 3-D device, and then
  // activate tone generator

  void play_sound(int x, int y)
  {
  // set 3-D device azimuth via MIDI controller 100
  //send_midi_control(chan. #, cont. #, cont. value)

       if      (x < 213) send_midi_control(2,100,0);
       else if (x > 426) send_midi_control(2,100,127);
       else              send_midi_control(2,100,64);

  // set 3-D device elevation via MIDI controller 101

       if      (y < 160) send_midi_control(2,101,0);
       else if (y > 320) send_midi_control(2,101,127);
       else              send_midi_control(2,101,64);

  // play tone generator middle C via MIDI note on

       send_midi_note_on(1,60,127);
  }
```

FIGURE 5.6. A C language program for interfacing a tone generator and a 3-D audio subsystem in response to a mouse click. The send_midi_xxx type functions call a lower–level function that transmits the proper byte sequence to the serial port for the particular command. send_midi_control sends MIDI channel number, controller number, and controller value; send_midi_note_on sends MIDI channel number, note number, and velocity (See Figures 5.4–5.5).

HEAD-TRACKED 3-D AUDIO FOR VIRTUAL ENVIRONMENTS

In virtual reality, the key to immersivity is directly tied to the success of the interface in providing realistic information at the effectors and the degree to which sensors are able to track human interaction. For this reason, **head-tracked virtual acoustic systems** represent the best approach to implementing sound, so that sound sources can remain fixed in virtual space independent of head movement, as they are in natural spatial hearing.

Crystal River Engineering

The Convolvotron, one of the first head-tracked virtual audio systems, was originally developed by Scott Foster and Elizabeth Wenzel for the NASA's groundbreaking VIEW (Virtual Interactive Environment Workstation) project, which helped launch virtual reality into the public eye in the late 1980s (Fisher, et al., 1986; Foley, 1987; Fisher, et al., 1988; Wenzel, Wightman, and Foster, 1988). The device is now available commercially from Crystal River Engineering, Palo Alto, California. Essentially, a Convolvotron is an integrated 3-D audio subsystem on two cards placed within an IBM PC-AT compatible computer (see Figure 5.7). The Convolvotron's spatialization functions can be run either through supplied software or via routines that can be called within a C language program by including a special library. Combined with a head tracker such as a Polhemus Isotrak, the device can spatialize up to four input sources at positions independent of the user's head position, while maintaining a sample rate of 50 kHz. Virtual sources can move about the listener, independently (as a function of sound movement software) and/or dependent upon the relative position of the head.

The Convolvotron uses 256-coefficient, minimum-phase HRTFs; 74 pairs of filters are accessed, from positions measured at every 15 degrees azimuth at five different elevations (−36, −18, 0, 18, 36, and 54 degrees). A four-way interpolation similar to that shown in Figure 4.38 is used to obtain a synthetic HRTF for an arbitrary virtual acoustic position; the interpolation is updated at a 33 Hz frame rate for smooth transitions in response to head motion. The device has a specified average delay of 32 msec for directional controls and audio through the system.

The device operates on blocks of source data 5 to 20 msec in length, before obtaining an additional block of samples at the inputs. At the beginning of each block, directional data is provided by the host computer for each source in use for specifying sound movement, based either on head-tracker and/or software-specified movement trajectories. The directional data consists of four weights and four pointers to the measured HRTF impulse responses stored within the Convolvotron. These weights and pointers are used to compute an interpolated response for the desired direction; the result is a binaural pair of

FIGURE 5.7. The Convolvotron signal-processing boards. Top: The modified Spectrum TMS320-C25 system board. Bottom: The "convolution engine" board.

impulse responses representing a synthetic HRTF. These impulse responses are convolved with the first block to provide left and right output signals. The total binaural output is formed by summing the filtered output from each source in use. To provide smooth movement transitions, the device computes extra samples at the end of each block. These samples are then faded into the next block to avoid artifacts such as clicks or jagged spatial transitions resulting from computation of virtual source movement or head motion.

The Convolvotron software consists of a library of programs that are run from a standard DOS operating system for various types of demos, tests, signal analyses, and the like. This flexibility allows the device to be run as an integrated 3-D audio system. But most users take advantage of the software library for integrating the Convolvotron's spatialization functions within the central "reality engine" program. In this way, global variables about the relative position of visual objects can be linked to virtual auditory program statements. For example, the library routine `ctron_locate_head` locates the listener's

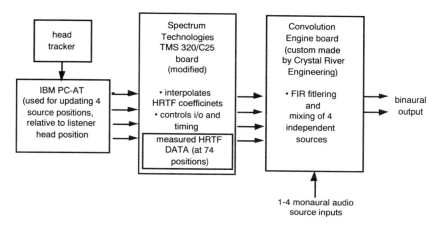

FIGURE 5.8. The Convolvotron, shown in block diagram form. In reality, one A/D and D/A is on the Spectrum board. *Adapted from Wenzel, et al., 1990.*

head position relative to the coordinates of the virtual world; `ctron_locate_source` does the same thing for one of the virtual source locations. The signal processing can also be tailored for the individual user; for instance, the function `ctron_model_head` allows a customized interaural axis to be specified (since head sizes differ) and adjusts the vertical offset representing the distance from the head's center point to the location of the head tracking sensor (usually mounted on the top of the headphone band).

The Convolvotron hardware consists of two boards, one termed the "Spectrum board," and the other the "convolution engine" card, that plug into the ISA or EISA bus slots of the PC-compatible computer (see Figure 5.8). The spectrum board is a modified TMS320-C25 system board, built by Spectrum Signal Processing, that includes a Texas Instruments TMS320 DSP chip and single A/D and D/A. The spectrum board functions primarily to perform the interpolation calculations for the convolution engine card, as well as performing input data buffering and clock signal generation. The modifications to the original Spectrum board are primarily expanded memory for the real-time control software. When the input buffers on the Spectrum board are loaded, they get passed to the second card for "background processing," i.e., FIR filtering. The "convolution engine" is a custom-made board that contains four INMOS A-100 cascadable DSPs, each containing 32 parallel 16-by-16 multipliers, allowing for a peak convolution speed of 320 million taps/sec. The convolution engine also holds the three additional A/Ds and a single D/A for audio input-output. This card receives the interpolated coefficients from the Spectrum board, along with sound input data buffers. Upon completion of the convolution sequence, the mixing of the left-right outputs from the four sources is performed and then output according to the Spectrum board's DSP clock.

Crystal River Engineering also manufactures a "little brother" and a "big brother" version of the Convolvotron. The Beachtron™ is a low-cost version that spatializes two sources in real-time and actually is a modified Turtle Beach Multisound™ multimedia card. The card itself works with a Motorola 56001 DSP rather than the A-100 and also contains an Emu Proteus™ synthesizer chip. In addition, it offers MIDI synthesizer and audio recording and playback capabilities.

A full-blown device known as the Acoustetron™ is intended for use in a multiuser, distributed virtual reality system. It contains its own PC and card cage that can be configured with up to eight Convolvotrons. It also contains special client-server software to provide audio spatialization for virtual reality engine computer hosts (e.g., a Silicon Graphics or Sun workstation). Crystal River Engineering also has developed software for the Convolvotron-Acoustetron for real-time spatialization of early reflections (Foster, Wenzel, and Taylor, 1991; CRE, 1993) called "The Acoustic Room Simulator." One can provide information from an image model program, as described in Chapter 4, to simulate and spatialize six early reflections, using 128-tap FIR HRTF filters, both in real-time and in response to head movement. The program also allows simple filtering for modeling frequency-dependent wall materials and for moving a source behind a wall using a "transmission loss model" implemented as a low-pass filter. The dynamic spatialization of early reflections is quite impressive in spite of the simplicity of the acoustical room model, especially when source and listener are on opposite sides of a modeled wall. In the future, **head-tracked auralization systems** using similar techniques may be particularly important for resolving front-back ambiguities present in current auralization simulations (see Begault, 1992c).

Focal Point

3-D audio systems were developed early on by Bo Gehring and Gehring Research Corporation (GRC) of Toronto. A 3-D sound application patent, "Attitude Indicator" (Gehring, 1988) describes several concepts for the use of virtual auditory cues in military cockpits. The basic concept was implemented into several binaural convolution devices intended primarily for military aerospace research. An early version, the AL-100, used a head-tracked, servo-positioned Kemar artificial head in a box lined with thirty-six loudspeakers (Doll, et al., 1986). The AL-204 (see Figure 5.9), was an early four-channel DSP-based convolution device that used HRTFs based on measurements of the KEMAR mannequin head (Figure 4.25) made at Northwestern University (cf. Kendall, Martens, and Wilde, 1990). This hardware was designed to provide auditory location coordinates and spatialized audio input from flight control and other onboard systems.

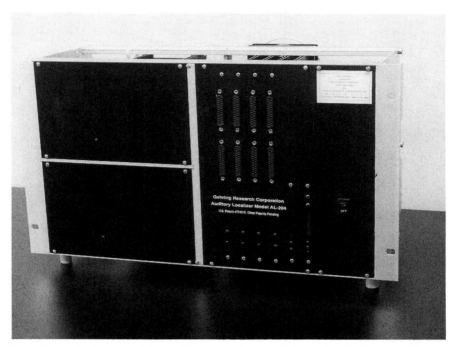

FIGURE 5.9 The AL-204 auditory localizer, a 3-D audio device capable of interpolation and four-channel spatialization, manufactured in 1988. *Courtesy of Bo Gehring.*

Gehring later entered the virtual reality market with a low-cost device known as the Focal Point™ system, a head-tracked virtual acoustic display for either the Macintosh or the PC. The Macintosh version used a modified Digidesign Audio Media™ card, which was one of the first widely available multimedia cards specifically for sound. It has been used by many third-party developers for their specific audio applications, since a Motorola 56001 DSP, stereo A/D–D/As, and a mic preamp are contained within one card, and a set of software routines allows direct-from-disc recording and playback to be accomplished relatively easily. The HRTFs used in Focal Point systems are proprietary.

A Focal Point system for the PC followed the Macintosh-based implementation. The Focal Point Macintosh card processes one sound source in real time within the host, while the PC card can process two sound sources; in both systems, one can spatialize additional sources by inserting additional cards. The software includes provision for spatializing sound resources or sound files directly from disk and has a very useful control panel that allows spatialization as a function of mouse position. The Macintosh version also works under Apple's MIDI Manager, which allows MIDI input to drive the spatialized

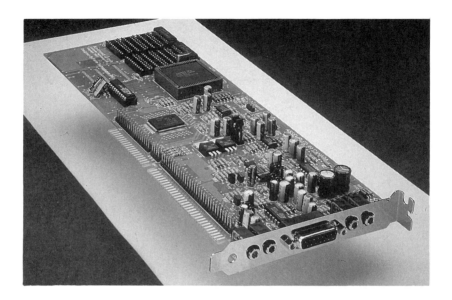

FIGURE 5.10. The Advanced Gravis UltraSound Sound card for PC compatibles. Featuring Focal Point Type 2 software for sound spatialization, it is the first but definitely not the last low-cost computer game card featuring 3-D audio. *Courtesy of Advanced Gravis Computer Technology, Ltd.*

position by connecting icons on the computer. The software contains a limited but optimized set of routines for initialization, setting of sound source position, and support for head tracking that can work independently of the CPU.

The Focal Point Type 2 software (patent pending), licensed by Advanced Gravis for use on its UltraSound™ multimedia sound card (see Figure 5.10) uses a non-real–time preprocessing utility to produce from three to six copies of an input source, typically at 90-degree intervals of azimuth along with elevated positions. Sounds are stored and played back from the computer hard disk or looped from sample memory. Spatial positioning on playback is done by mixing particular combinations of these prefiltered sounds using the Focal Point Type 2 driver algorithm. For instance, to move from left 90 degrees to left 45 degrees, one makes a transisition, similar to amplitude panning, between the left 90-degree and the 0-degree position.

A Virtual Reality Application: NASA's VIEW System

The implementation of the Convolvotron 3-D audio device within NASA's VIEW system gives an example of how distributed 3-D audio devices and tone generators are integrated within a dedicated virtual reality application. Wenzel, *et al.* (1990) created a distributed audio system as part of a virtual environment for providing symbolic acoustic signals that are matched to particular actions in terms of the semantic content of the action taken. The sounds are not only spatialized in 3-D dimensions for conveying situational awareness information but are also interactively manipulated in terms of pitch, loudness and timbre.

Figure 5.11 shows the hardware configuration for the auditory display. The reality engine at the time consisted of a Hewlett Packard HP900/835 computer. The RS-232 port was used to drive a Digital Equipment DECTALK™ speech synthesizer, a Convolvotron, and two Ensoniq ESQ-M™ tone generators via an RS-232-MIDI converter. In the configuration, the Convolvotron receives data over the RS-232 port regarding the head-tracker orientation and action of a VPL Data Glove™, and spatializes the output of one of the two synthesizers. A standard audio mixer is used to sum the outputs of the synthesizer, Convolvotron, and speech synthesizer.

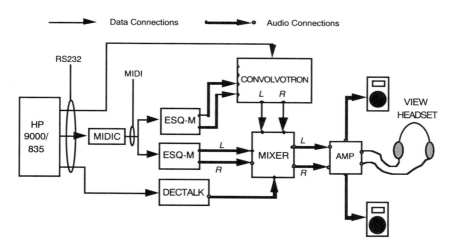

FIGURE 5.11. Implementation of the Convolvotron into the VIEW laboratory at NASA Ames Research Center (about 1987–1988). MIDIC is an RS-232 to MIDI converter. *Adapted from Wenzel,* et al., *1990.*

The Ensoniq ESQ-M tone generators were set to receive a variety of MIDI control information. Oscillator frequency, filter cut-off, and other parameters that are the building blocks of auditory icons can be mapped from the action of sensors that are input to the reality engine. While this type of configuration can be a formidable chore under normal circumstances, the VIEW audio subsystem was configured via specialized software (the auditory cue editor, ACE, developed by Phil Stone) that eased formation of the auditory symbologies and their connections to virtual world. In particular, the ability to do this off-line (in a stand-alone manner) from the reality engine is a practical consideration where availability of resources for development are frequently at a premium.

As already suggested earlier in this chapter, audio in any virtual reality configuration can provide an important source of feedback. For instance, the VPL Data Glove works by analyzing a combination of detected finger positions and then using the software to match them to sets of predefined gestures (e.g., a peace sign, pointing with the index finger, squeezing hand as a fist). It can be tricky at first to adapt one's hand actions to recognized gestures because individual variation in hand size and movement must match a generalized gesture. But if auditory feedback is supplied upon successful recognition, it aids the user in knowing when the action has been completed successfully. At Stanford Research Institute (SRI), Tom Piantanida and his colleagues supply auditory feedback when a fist gesture is recognized by playing a "squish" noise from a digital sampler that sounds similar to a wet washcloth being squeezed in the hand. This response is a more literal, or "representational," use of sound and might be termed an auditory icon since it is representative of an everyday sound (Gaver, 1986). By contrast, the VIEW project used a combination of two synthesized tones (e.g., the musical notes C and D) to give the user "glove state" feedback: An ascending sequence indicated a glove opening, and the reversed sequence indicated the glove closing. This type of "abstract" pitch sequence to which meaning is attached might be referred to as an nonrepresentational **earcon** (Blattner, Sumikawa, and Greenberg, 1989).

A common use of representative aural cues in virtual reality is as a means of replacing tactile sensation (haptic feedback) that would normally occur in a real environment. In virtual reality, one can walk through walls and objects; but by activating auditory information in the form of a bumping or crashing sound, the user receives feedback that a particular physical boundary has been violated. But after a while, using a simple crash for all encounters with different types of boundaries becomes increasingly unsatisfying, because it lacks complexity in response to human interaction. While the VIEW system had general application to many sorts of virtual environment scenarios and certainly could make use of these "representational" auditory cues, its use of sound for interaction in a **telepresence** context was vastly more sophisticated.

An example is the auditory "force-reflection" display used to substitute for a *range* of haptic (tactile) sensation involved in the teleoperated placement of a circuit card. Wearing a helmet-mounted display and data glove, the user was faced with the task of guiding a robot arm by watching the action within the virtual environment. Once the user's arm and hand were coupled with the robot equivalent and had grabbed the circuit card, the task was to correctly guide the card into a slot, a rather difficult undertaking with only visual feedback, given the latencies and complexity of visual displays of the time. In this scenario, **continuous variation** of sound parameters was used. If the card was being placed incorrectly, a basic soft, simple tone got louder, brighter, and more complex (via frequency modulation) the harder the user pushed to discourage damage to the misaligned card. This sound was accomplished by sending appropriate MIDI information on note volume and frequency modulation to the tone generator.

Another form of auditory feedback described by Wenzel, *et al.* (1990) was the use of **beat frequencies** between two tones to continuously inform the listener about the distance between the card and the slot. You may have noticed that a guitar being tuned produces a beating effect when two strings are close but not exactly in tune. As two identically tuned strings are slowly brought out of tune, the frequency of the beating increases. These beats are actually the perceived amplitude modulation of the summed waves, caused by the auditory system's inability to separate two frequencies on the basilar membrane smaller than the limit of discrimination (see Roederer, 1979). Using the beating as an acoustic range finder, the user could manipulate the card until it was successfully in the slot; the closer the card was to the slot, the closer the two tones came to the same frequency, with a corresponding reduction in the beat frequency. When the card was installed, the tone stopped, the voice synthesizer would say "task completed," and a success melody was played (a quickly ascending scale, similar to the sound of a Macintosh starting up).

COMPUTER WORKSTATION 3-D AUDIO

A Generic Example

A simple audio equivalent to the graphic user interface (GUI) as described previously is only marginally useful. The concept of a computer workstation as not only a location for editing documents but also for **communicating** with databases and persons within the office adds a new dimension to the possibilities for 3-D sound. For instance, telephone communication can be handled though a computer via functions such as message recording and dialing. If the audio signal is brought into the computer and then spatialized, the communication source can be usefully arranged in virtual space. This feature becomes particularly important in multiconversation **teleconferencing**, where video images of the participants can be broadcast on the screen. Spatial audio teleconferencing has been described as an **audio window** system by Ludwig, Pincever, and Cohen (1990).

Consider all of the possible audio inputs to a workstation user, not only teleconferencing environments, but all types of sonic input. All types of sonic input could be directionalized to a specific location, controlled by the user. Furthermore, it is not necessary to place these audio sources in their corresponding visual locations; the audio spatial mapping can correspond to a **prioritization** scheme. For telephone calls, spatialization of incoming rings will become interesting when the caller can be identified. In such a scenario, calls from subordinates could be signaled by a phone ringing from the rear; a call from home could always ring from front right; and the intercom from the boss could be from front and center. In this case, spatial location informs the listener as to the prioritization for answering calls—whether or not to put someone on hold, or to activate a phone mail system.

Other types of input that could be spatialized to a headphone-wearing workstation user include building fire alarms, signals for sonification of the state of a particular computer process of machinery in another room, an intercom from a child's crib, as well as the more familiar "system beeps" that provide feedback from software such as word processors. And there's no reason why one's favorite CD cannot be placed in the background as well.

Figure 5.12 shows a hypothetical GUI for arranging these various sound inputs in virtual auditory space in a manner configurable by each individual listener. The basic concept is to represent each form of auditory input iconically on the screen and then allow manipulation of the distance and relative azimuth of each source. The computer interprets the relative location of the icons on the screen as commands to a multiinput, 3-D sound spatialization subsystem.

In noisy environments, the 3-D audio workstation system could also include **active noise cancellation**. A microphone placed on the outside of the headphones is used with a simple phase-inversion circuit to mix in a 180-degree inverted version of the noise with the desired signal at the headset. In reality, active noise cancellation headphones cannot completely eliminate all noises; most models provide about a 10 dB reduction in steady-state sound frequencies below 1 kHz. But when no signal is applied to the headphones except the inverse noise signal, it is surprising how well active noise cancellation headphones work in a typical office environment. In informal experiments by the author using lightweight Sennheiser HDC-450 NoiseGard® mobile headsets, the sound of HVAC systems, whirring computer fans, and outdoor traffic is noticeably reduced. The potential for using active noise cancellation in workstation 3-D audio systems seems great.

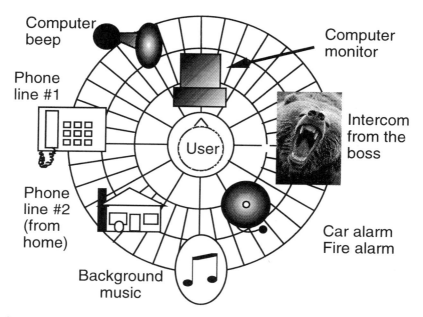

FIGURE 5.12. Layout for a hypothetical GUI for arranging a set of incoming sounds. By resizing the icons, volume could be adjusted independently of perceived distance. *Graphic assistance by Denise Begault.*

Audio Windows

The concept of **audio windows** has been explored extensively at the Human Interface Laboratory of Nippon Telegraph and Telephone Corporation by Cohen and his associates (Cohen and Koizumi, 1991; Cohen, 1993). They have developed an experimental glove-based system called "handysound" and a more viable interface called MAW (Multidimensional Audio Windows) that uses 3-D sound to control the apparent azimuthal location of conference participants. A user of the MAW system sees a display of circular icons with pictures of each participant from a bird's eye perspective, arranged about a virtual desk. The visual arrangement of icons translates into instructions for a 3-D audio subsystem.

The gain and azimuth heard for each icon is linked to the relative position on the graphic display; a mouse can be used to make sources louder by resizing the icon. MAW allows teleconference participants to wander around in the virtual room and to focus their voice toward particular conference participants by using a binaural HRTF to process each virtual source. Distance cues also are modeled in the system, and one can leapfrog between teleconference sessions or be at several simultaneously simply by cutting and pasting one's personal graphic icon to different environments, each represented by a separate window. One can also move to a "private conference" by moving the icons of the participants to a special region on the computer screen, symbolized as a separate room.

Audio GUIs for the Blind

Burgess (1992) and Crispien and Petrie (1993) have described interesting applications of spatial sound for allowing blind persons to navigate within a GUI. Crispien and Petrie's concept is applicable to either a headphone-based 3-D sound system or a multiple loudspeaker system. Auditory icons were interfaced with a Microsoft® Windows™ operating system using the icons to respond to various GUI interactions (see Table 5.2). These sounds were then spatialized according to the position on the computer screen by a two-dimensional auditory display (azimuth and elevation). Because GUIs are not friendly operating systems for blind persons, development of spatialized acoustic cues might have great potential. But presently there are no studies comparing blind people's non-GUI system performance versus a 3-D audio cued, GUI system where subjects are given equivalent periods of training and practice on both systems.

Interactions	Auditory icons
mouse tracking	steps walking
window pop-up	door open
window moving	scratching
window sizing	elastic band
buttons	switches
menu pop-up	window shade

TABLE 5.2. Some of the spatialized auditory icons used by Crispien and Petrie (1993) in their GUIB (Graphical User Interface for the Blind) system. Interactions are actions taken by the user on the GUI; auditory icons, a term coined by Gaver (1986) are sounded in response to actions on the screen and then spatialized to the appropriate position. Compare to Gaver's (1989) *The Sonic Finder: An Interface That Uses Auditory Icons.*

RECORDING, BROADCASTING , AND
ENTERTAINMENT APPLICATIONS

Loudspeakers and Cross-talk Cancellation

Many people want to control 3-D sound imagery over **loudspeakers** as opposed to headphones (see the section in Chapter 1, "Surround versus 3-D Sound"). Most have experienced "supernormal" spatial experiences with normal stereo recordings played over stereo loudspeakers, and there are countless methods to produce exciting spatial effects over loudspeakers that require nothing of current 3-D sound technology. And there have been and probably always will be add-on devices for creating spatial effects either in the recording studio or in the home (one of the oldest tricks involves wiring a third in-between speaker out-of-phase from the stereo amplifier leads). The problem is that it's difficult to harness 3-D spatial imagery over loudspeakers in such a way that the imagery can be transported to a number of listeners in a predictable or even meaningful manner.

There are three reasons why 3-D sound over loudspeakers is problematic, compared to headphone listening.

1. The environmental context of the listening space, in the form of reflected energy, will always be superimposed upon the incoming signal to the eardrums, often with degradative or unpredictable effects. This distortion is especially true for strong early reflections within 10–15 msec, although these can be minimized by listening in a relatively acoustically damped room or in an anechoic chamber.

2. It is impossible to predict the position of the listener or of the speakers in any given situation and impossible to compensate for multiple listeners. One can go to extremes to put people in what is known as the **sweet spot**, best defined as the location closest to where the original recording was mixed by the audio engineer. At one audio product show, a demonstration room for a 3-D sound system had four small chairs arranged in a row on a central line between the speakers but so closely that one's chin was practically touching the back of the next person's head: an unlikely scenario in an actual application. In the area of music reproduction, the culture surrounding home stereo listening seldom favors a single listener—the audiophile, who actively listens while facing the loudspeakers. More often than not, loudspeakers are heard as background music, and/or by multiple persons, while watching a television monitor.

3. Unlike the sound from headphones, the acoustic signals arriving at each ear are a mix of the signals from the two loudspeakers, as shown in Figure 5.13. 3-D sound effects depend on being able to predict the spectral filtering occurring at each ear, which can be predicted reasonably well for diffuse-field corrected headphones (see the section in Chapter 4, "Equalization of HRTFs"; unfortunately, the coupling between the headphones and individual pinnae

cannot be predicted). But each loudspeaker is heard by both ears; this cross-talk will affect the spectral balance and interaural differences significantly. The attenuated and delayed signal from the opposite channel is known as the **cross-talk** signal, since it crosses to the opposite ear. If one plays a 3-D sound recording mixed for headphones over loudspeakers, the two channel signals that were discrete at each ear become: left + right speaker signal, attenuated and delayed, at the left ear; and right + left signal, attenuated and delayed, at the right ear.

The solution for this dilemma, known as **cross-talk cancellation,** was originally proposed by Schroeder and Atal (1963). If one can ignore the effects of the listening environment, and especially if the position of the head relative to the speakers is known, then one can add to the left binaural signal an inverse (180 degrees out of phase) version of the cross-talk signal, delayed by the amount of time it takes to reach the opposite ear. In theory, the inverse waveform cancels the cross-talk, since the addition of two waveforms 180 degrees out of phase results in a perfectly canceled signal, as shown in Figure 5.14.

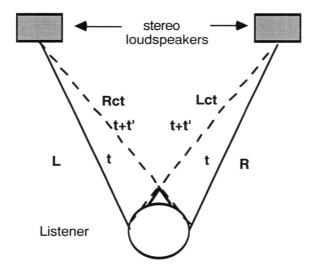

FIGURE 5.13. Illustration of cross-talk cancellation theory. Consider a symmetrically placed listener and two loudspeakers. The cross-talk signal paths describe how the left speaker will be heard at the right ear by way of the path *Rct,* and the right speaker at the left ear via path *Lct.* The direct signals will have an overall time delay *t* and the cross-talk signals will have an overall time delay *t* + *t'.* Using cross-talk cancellation techniques, one can eliminate the *Lct* path by mixing a 180-degree phase-inverted and *t'*-delayed version of the *Lct* signal along with the *L* signal, and similarly for *Rct.*

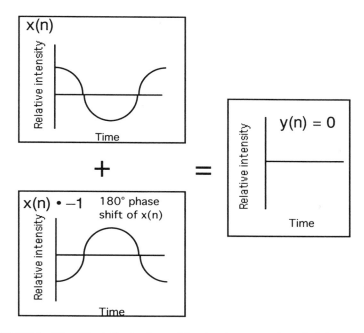

FIGURE 5.14. Simplified illustration of the basic concept underlying cross-talk and active noise cancellation theories. If the undesired signal is the waveform *x(n)* at the top, a simple phase-inverting circuit can produce the 180° out of phase version below. If the two signals are mixed electronically, perfect cancellation occurs (summed signal at right with 0 intensity). In actual practice the 180° phase-shifted waveform will also be time delayed and possibly HRTF-filtered.

A brief analysis reveals the importance of the fact that the cross-talk signal is effectively filtered by the HRTF before reaching the opposite ear. So the solution would be to prefilter the cross-talk signal by an HRTF—but which one? It is difficult to know the angle of incidence of loudspeakers in an arbitrary setting, and as a listener moves away from a set of fixed loudspeakers, the angle of incidence narrows, which in turn changes the HRTF necessary for use in the cross-talk calculation factor.

How much head movement is allowable? Assume that the average head has ears separated approximately 9 inches and that the speed of sound is 344 meters per second. A 2,000 Hz wavefront would then be about seven inches long, and a cross-talk canceled version the same length but out of phase 180 degrees. As cross-talk cancellation occurs in air, moving the head laterally side to side three-and-a-half inches to the left would cause the 180-degree out-of-phase cross-talk signal to flip back into 0-degree phase at the left ear, but the direct signal would *not* be offset by this amount. The moral is that normal head movements cause summed relationships that are different from the desired total cancellation.

Hence, without a fixed head (impossible in any normal auditioning), only the lowest octaves of musical material will be successfully cross-talk canceled.

Some progress is being made in the area of making 3-D sound a satisfying experience over loudspeakers (see, for example, Cooper and Bauck, 1989; Griesinger, 1989; and Kendall, Martens, and Wilde, 1990). There is a commercially available device that allows remote adjustment for speaker angle and listening distance and that incorporates cross-talk cancellation of the cross-talk signal (see Griesinger, 1989). But despite the best efforts, cross-talk cancellation probably requires further development for most contexts; the 3-D sound system needs input about the details of the listening space and the location of the loudspeakers and listener to be effective. Perhaps someday someone will figure out how to make head trackers good enough (and light enough, without cables) so that a computer could calculate head position relative to several sets of loudspeakers and provide continuously varying cross-talk cancellation signal processing.

Binaural Processors for the Recording Studio

The application of 3-D sound to the entertainment world is where virtual reality and money meet, in that research funding matches the considerable number of units sold. While these are not direct implementations of 3-D sound into virtual reality or multimedia, technological innovations cross-feed between the entertainment and professional recording industries and more exploratory, scientific uses of 3-D sound.

In the end of his review on spatial hearing, Blauert (1983) mentioned the concept of a **binaural mixing console**, a 3-D sound processor built into each input strip of a commercial recording mix-down console. Normally, the art of mixing a commercial recording for an unknown medium (different loudspeaker or headphone configurations) requires that the spatial imagery be highly **malleable**. Most professional recording engineers have the ability to hear their mix independent of the differences in tonal coloration (timbre) between different speakers and between different listening environments. Even within a single control room, the virtual sound source positions resulting from a stereo mix will be different—in the absolute sense of how many degrees separation there is between the hi-hat and the keyboard track—when switching between the speakers placed on top of the console and those mounted on the wall. The key here is that the **relative placement** of sources remains constant with amplitude panning. 3-D sound, on the other hand involves frequency-dependent time and amplitude differences that are very fragile; in other words, the relative placement of imagery doesn't transfer between different speakers or rooms very well. Nevertheless, many engineers and manufacturers are using 3-D sound techniques for mix-down of commercial recordings and feature 3-D sound as a special feature of the recording.

AKG Corporation's Binaural Consoles

An earlier implementation of a binaural mixing consoles was found on the CAP 340M Creative Audio Processor™, developed by AKG Corporation (Akustiche und KinoGeräte GmbH, Vienna) in 1987 as a prototype "blank slate virtual recording console." Its computing power of 340 megaflops was in line with the fastest supercomputers of the time. Approximately the size of a small refrigerator, the CAP 340M consists of several rack-mount units and a computer user interface that includes a "virtual mixing console," with overhead and side perspective pictures of a head replacing the normal panpots for specifying a position for each input source 3-D space. Several HRTF sets and custom-designed filters could be used with the system, and room simulation was also available. Provision was made for using several types of microphones (for instance, close up spot microphones and a more distant dummy head mic, in front of an orchestra) and then sorting out their directivity patterns and time delays to allow mix-down compatibility and simulation of acoustic perspective.

The technology contained in the CAP 340M processor was later taken into AKG's Audiosphere BAP™ 1000 device, introduced in Europe in the early 1990s (Persterer, 1991). This is a more limited, yet tailor-made application of virtual acoustics to a specific problem. The ideal monitoring system for any professional audio broadcasting or recording system would include loudspeakers and an environment that directly matches that of the listener. This, of course, is difficult to predict for the small percentage of persons listening over headphones and impossible for loudspeaker listeners. Frequently, two or three generalized speaker systems are used by recording engineers to monitor the difference between, for example, a "generic high-fidelity system" and a "generic low-fidelity" system (e.g., automobile or portable stereo speakers). Often, an engineer must switch between studio control rooms in the middle of a mixing project; and in live broadcasting from mobile units (a more common practice in Europe than in North America), the engineer often has no choice but to use headphones.

The BAP 1000 is a virtual acoustic **monitoring** system, in that 3-D sound techniques are not imposed for any other purpose but to allow simulation of a pair of stereo loudspeakers in an ideal recording studio. The device filters the output from a mixing console using transfer functions that represents the HRTF impulse responses of loudspeakers in an ideal control room, at left and right 30 degrees azimuth. The control room early reflections are also HRTF filtered. One can then listen over a set of system-equalized AKG K-1000 headphones connected to the BAP 1000 and have an externalized, loudspeaker experience. Users can switch between nine different HRTF sets, which the manufacturer claims represent 90 to 95 percent of the population. In the future, recording engineers or home listeners could have their own HRTFs measured in their

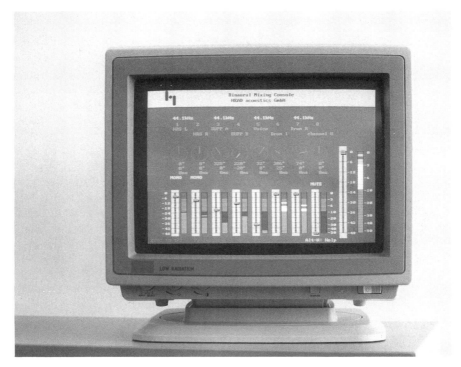

FIGURE 5.15. The interface for the BMC binaural mixing console manufactured by HEAD acoustics. *Courtesy of Sonic Perceptions–HEAD acoustics.*

favorite control rooms and then have the data stored on a customized ROM card. This data base potentially would allow switching between virtual control rooms representing the best studio environments while conducting work in a much less expensive facility.

HEAD acoustics

HEAD acoustics (Hezongrath, Germany; distributed by Sonic Perceptions, Norwalk, Connecticut) manufactures a range of high-end measurement, recording, and headphone playback devices related to 3-D sound. Their system includes not only the HRS II dummy head (similar to the head shown in Figure 4.27), but also an eight-channel binaural mixing console that uses HRTF filtering to simulate the effects of the dummy head on monaural inputs. The binaural mixing console features a graphic monitor display of mixing console faders for each input and spatial panpot controls (see Figure 5.15). Each input can be attenuated, delayed up to 300 msec and then HRTF filtered to an arbitrary position.

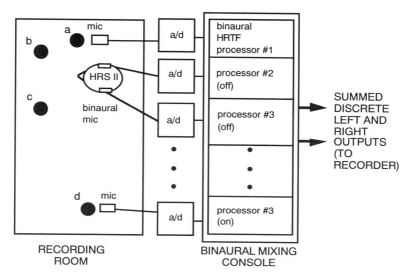

FIGURE 5.16. Use of a binaural mixing console with a dummy head for recording. The filled circles in the recording room labeled *a-d* are individual sound sources. Two spot mics (microphones placed very close to the sound source) are used on sources *a* and *d*; all four sources and the environmental context are picked up by the dummy head (here, the HEAD acoustics HRS II). The binaural mixing console (HEAD acoustics BMC) takes eight inputs and produces a summed stereo pair; inputs can be individually attenuated, time delayed, and HRTF filtered. The processors are shown as turned off for the dummy head inputs (#2 and #3 inputs on the binaural mixing console) since they are already HRTF-processed by the dummy head. *Courtesy of Sonic Perceptions-HEAD acoustics (adapted).*

The HRTFs used in the HEAD acoustics BMC are based on the HRS II dummy head's "structurally averaged" pinnae, developed by Genuit (1984). Internally, the BMC contains measured HRTFs at every two degrees of azimuth and five degrees of elevation, causing the interpolation problems mentioned in Chapter 4 to be minimized. One could use a binaural mixing console such as this to spatially process each track of a multitrack tape source into a final stereo mix-down. Alternatively, for several instruments recorded live, one can mix in a combination of dummy head and electronically HRTF-filtered sources, as shown in Figure 5.16, since the dummy head and the electronic HRTFs will be compatible. A single input can also be routed to several processing units; the second through *n* inputs can be set to delays, attenuations, and HRTF values corresponding to typical timings and levels for the first few early reflections. This procedure tends to vastly improve loudspeaker reproduction in general, and, concurrent with the data shown previously in Figure 3.9, aids in the externalization of headphone imagery (see Plenge, 1974; Sakamoto, Gotoh, and

Kimura, 1976; Begault, 1992a); however, the number of available inputs is correspondingly reduced.

Roland Corporation

One of the first viable 3-D audio mixing consoles aimed primarily at the commercial music industry was called the Roland Sound Space™ (RSS) processing system; it has been featured on several commercial recordings since about 1990. The system consists of a mixing console user interface (the SSC-8004 Sound Space Controller Unit) and two rack mounted units, one for spatial processing and cross-talk cancellation circuitry, the other for analog-digital-analog conversion (see Figure 5.17).

The user interface of the RSS is notable in that it consists of a panel of rotary knobs for adjusting azimuth and elevation independently. Up to four input channels can be independently placed in virtual space. While the RSS was primarily intended for loudspeaker playback, it can be successfully used for

FIGURE 5.17. The first generation: the Roland Sound Space (RSS) processor. The user interface is in the foreground: dials are used to set azimuth and elevation separately. The processing unit and the A/D and D/A conversion unit are shown in the background.

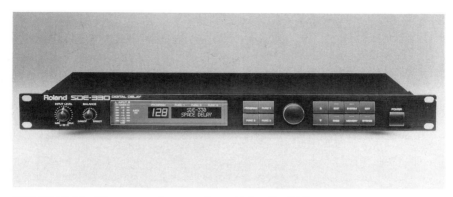

FIGURE 5.18. The second generation: Roland's SDE-330 space delay device. This is a self-contained, single rack-mount sized unit.

headphones by bypassing the cross-talk cancellation circuitry (Jacobson, 1991). MIDI input can be used with the SRS to control spatial position in live performance as well, and the device could be integrated within a virtual reality system to control 3-D sound imagery via MIDI control. Little information is publicly available on the device's internal workings, the HRTFs used, or the interpolation factors that are included.

In 1993, Roland introduced its SDE-330 "dimensional space delay" and SRV-330 "dimensional space reverb" signal processors. Unlike the SRS, which was rather large, intended primarily for studio applications, and priced over $40,000, the SDE/SRV-330 units are single rack-mount units, intended for either live or studio use, and are considerably less expensive. They contain ultra-high–speed DSPs (67.74 and 33.38 megaflops, respectively) and perform 30-bit internal processing and 16-bit conversion at the A/D and D/A converter at a 44.1-kHz sampling rate.

The SRV-330 can provide forty early reflections at twenty different spatial locations, as well as the standard ensemble of late reverberation effects (e.g., "small room," "plate"). The SDE-330, shown in Figure 5.18, is designed to allow eight delays to be placed at arbitrary positions on the horizontal azimuth and can combine pitch shift, feedback, and other standard delay unit functions with the internal 3-D sound processing. For instance, a direct sound from a voice can be mixed with four delayed, pitch-shifted versions that are placed at different locations around the listener. The reason for applying 3-D sound to these otherwise conventional functions is described in the product literature as follows: "The SD-330 can therefore produce a degree of spacy delay effects that other chorus effects on the market cannot." In other words, 3-D sound techniques, taken from the original RSS device, are used to create new interest in the competitive, high-volume market of relatively inexpensive effects units.

HRTF–Pseudo-Stereophony

One application of 3-D sound for loudspeakers developed by the author, **HRTF–pseudo-stereophony (HRTF–PS)**, is described in this section (Begault, 1992d). In professional audio, "pseudo-stereophony" describes a family of techniques that allow the derivation of two channels of sound from a monaurally recorded, one-channel source, usually with the object of creating the perceptual illusion that the sound was originally recorded in stereo. The techniques all involve a process whereby the two channels of output are derived according to a decorrelation process; i.e., the input is processed to create two noncoherent signals. The signal-processing techniques for achieving pseudo-stereophony have existed for at least forty years; a review can be found in Blauert (1983). Specific imaging is not the actual goal of pseudo-stereo; rather, an increased sense of image width and auditory spaciousness is important. It turns out that virtual acoustic techniques can be used for this application.

Recent interest in pseudo-stereophony was spurred by the development of television sets capable of receiving stereophonic sound in the 1980s. Although there are increasing numbers of televisions with stereo reception capability, a large number of broadcasts remain monaural. Therefore, the main application of pseudo-stereophony for television is to process monaural broadcasts to utilize the potential of the two built-in loudspeakers. The goal with the HRTF–PS circuit was to provide a stereo sensation, by imitating the decorrelation effects of a binaural micing situation with a simple DSP process. It is impossible to actually place virtual images at specific locations with this or any similar type of pseudo-stereo processor.

Most pseudo-stereophonic devices were designed for home or professional audio systems for postprocessing of monaural music recordings. At one time, the music industry infuriated many audiophiles by rereleasing classic monaural recordings of historical interest only in "electronically enhanced for stereo" versions; this trend seems to be waning. But there are important differences between music and television broadcast applications. For television, the pseudo-stereophonic process must be insensitive to the fact that television audio signals have a compressed dynamic range of 30 to 40 dB (compared to the approximately 65 dB dynamic range of an LP and the maximum 96 dB dynamic range of 16-bit DAC). Another important difference is in the type of signals transmitted. Home stereo listening is centered around musical material, while television audio material consists of both speech and music. Stereo systems and stereo televisions usually have very different speaker configurations. Some televisions allow the speakers to be angled inward or outward, and most systems allow the audio signals to be passed directly to an existent home stereo system.

An additional requirement for a television-based 3-D audio device is that the listening position relative to the loudspeakers not be critical. Thus the requirement for accurate perception of spectral imagery must be relaxed.

Televisions are often viewed and listened to by groups of people, only one of whom could be centrally located between the two speakers. Also, the distance from the set can be highly variable. Because some listeners will hear primarily only one speaker, the signal processing for a given channel cannot radically alter the timbre so that it is disturbing. Some earlier pseudo-stereo techniques suffered from this problem, particularly "complimentary comb-filtering," where each speaker had differentiated notched frequency responses (similar to Figure 4.48, except one channel's resonant center frequencies were alternated with the other channel's resonant center frequencies).

A final issue concerns the user interface. Audiophiles, the usual consumers of technologically sophisticated devices for home stereo systems, desire control over signal-processing parameters accompanied by visual feedback (i.e., the buttons, buzzers, and bells). The device in this case is an outboard processor purchased to augment an existing system. With a stereo television set, the consumer, who more than likely is *not* an audiophile, desires to buy an integrated product and may or may not desire interaction with the details of signal processing. As a consequence, the audio signal must be successfully processed into stereo without need of a user interface with variable controls. At best, any control should either be accomplished automatically or should exist as an optional feature for the consumer.

The HRTF–PS technique processes monaural sound according to a specialized model of a listener within an enclosure that has a very unnatural form of reflected sound. In its current implementation, only four early reflections are modeled. These would normally represent the first four single-bounce reflections found in the impulse response of the reverberation chamber; however, in this case the reflections are actually *louder* than the direct sound. This increase is necessary to accomplish the two-channel decorrelation within the available dynamic range.

A heuristic explanation of the HRTF–PS system is as follows: A monaural television speaker is placed in an anechoic chamber directly in front of a listener or a dummy head, as shown in Figure 5.19. In addition, four additional loudspeakers from the television are placed in an arrangement around the listener, at 90, 120, 240, and 270 degrees azimuth, 0 degrees elevation. These four additional loudspeakers are equipped with variable delay units and have an amplitude level that is 6 dB higher than the speaker facing the listener. Probe microphones are placed inside the ears of the listener or the mics of a dummy head are used; the pinnae match the average shape and transfer function of a number of listeners. The two-channel signal received at these microphones is played back through the speakers of the stereo television.

The angle of incidence of the reflections in relationship to the listener is an important factor. In the heuristic description of the model, four early reflections arrive from the sides and rear of the listener and are picked up at microphones

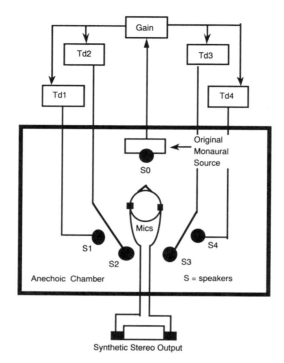

FIGURE 5.19. Diagram of the heuristic model for HRTF–PS (head-related transfer function pseudo-stereophony). *Adapted from Begault (1992d).*

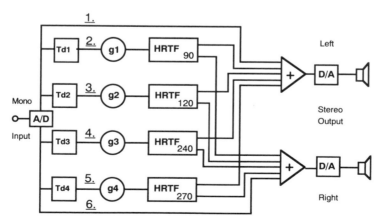

FIGURE 5.20. A system diagram for HRTF–PS, corresponding to the heuristic model shown in Figure 5.19. *Adapted from Begault (1992d).*

in the ear. Figure 5.20 shows the equivalent system diagram, which uses the linear-phase synthetic HRTFs described in Chapter 4. The time delays are set to typical values for early reflections, between 10 and 40 msec that correspond to larger or smaller versions of the "heuristic room" of Figure 5.19. The settings for the delay times are too great when speech reaches the "echo disturbance" point illustrated previously in Figure 2.5, corresponding to ITDs so large that two sounds are perceived (see also Haas, 1972). For this reason, a consumer implementation of HRTF–PS might have a four-position selector switch that would allow the time delay spread between Td1-4 to be increased or decreased. A more sophisticated circuit would automatically adjust the spread by predicting the source of the input material, i.e., whether or not the input had speech-like characteristics.

VOICE COMMUNICATION SYSTEMS

Binaural Advantages for Speech Intelligibility

The relationship between binaural hearing and the development of improved communication systems has been understood for over forty-five years (see reviews in Blauert, 1983; Zurek, 1993). As opposed to monotic (one ear) listening—the typical situation in communications operations—binaural listening allows a listener to use **head-shadow** and **binaural interaction** advantages simultaneously (Zurek, 1993). The head-shadow advantage is an acoustical phenomenon caused by the interaural level differences that occur when a sound moves closer to one ear relative to the other. Because of the diffraction of lower frequencies around the head from the near ear to the far ear, only frequencies above approximately 1.5 kHz are shadowed in this way. The binaural interaction advantage is a psychoacoustic phenomenon caused by the auditory system's comparison of binaurally-received signals (Levitt and Rabiner, 1967; Zurek, 1993).

Many studies have focused on binaural advantages both for detecting a signal against noise (the **binaural masking level difference**, or **BMLD**) and for improving speech intelligibility (the **binaural intelligibility level difference**, or **BILD**). Studies of BMLDs and BILDs involve manipulation of signal-processing variables affecting either signal, noise, or both. The manipulation can involve phase inversion, time delay, and/or filtering. The use of **speech babble** as a noise source has been used in several studies investigating binaural hearing for communication systems contexts (e.g., Pollack and Pickett, 1958).

Recently, speech intelligibility studies by Bronkhorst and Plomp (1988, 1992) have used a mannequin head to impose the filtering effects of the HRTF on both signal and noise sources. Their results show a 6 to 10 dB advantage with the signal at 0-degrees azimuth and speech-spectrum noise moved off axis compared to the control condition where both speech and noise originated from 0-degrees azimuth. According to a model proposed by Zurek (1993), the *average* binaural advantage is around 5 dB, with head shadowing contributing about 3 dB and binaural-interaction about 2 dB.

Another advantage for binaural speech reception relates to the ability to switch voluntarily between multiple channels, or streams, of information (Deutsch, 1983; Bregman, 1990). The improvement in the detection of a desired speech signal against multiple speakers is commonly referred to as the **cocktail party effect** (Cherry, 1953; Cherry and Taylor, 1954). This effect is explained by Bregman (1990) as a form of **auditory stream segregation**. Consider the experience of listening to music where multiple voices play against one another, as in a vocal chorale written by J. S. Bach. One can either listen to the entire "sound mass" at once or listen to and follow each individual voice. The binaural processing that allows one to "stream" each voice of a multivoice texture is not yet completely understood among psychoacousticians. Another effect related to the cocktail party paradigm is as follows: you can be actively engaged in a conversation with someone, but almost involuntarily you'll notice if your name comes up in another conversation. Both streaming and this type of subconscious monitoring of background noise are more difficult to do with one-ear listening. Next time you're in a noisy restaurant having a conversation with someone, try experimenting with these effects by closing one ear and noticing how much more difficult it is to understand speech conversation.

The Ames Spatial Auditory Display

Certain radio communication personnel, including emergency operators, military helicopter pilots, and the space shuttle launch personnel at NASA's John F. Kennedy Space Center (KSC), are almost daily faced with multiple channel listening requirements (Begault, 1993a). The current practice at KSC is to use single earpiece headsets, connected to a radio device capable of monitoring either four or eight separate channels. This situation is similar to the working conditions of police and fire service 911 operators, who use a combination of single earpiece headsets, telephones, and loudspeakers. A two-channel, general purpose head-tracked device specifically designed for fighter aircraft had been developed previously at Wright-Patterson Air Force Base and had been used with some success for improving speech intelligibility (Ericson, et al., 1993).

A four-channel spatial auditory display was developed by the author specifically for application to multiple-channel speech communication systems (Begault and Erbe, 1993). A previously specified design (Begault and Wenzel, 1990; Begault, 1992e) was used to fabricate a prototype device to improve intelligibility within KSC and similar communications contexts. The **Ames Spatial Auditory Display (ASAD)** places four communication channels in

FIGURE 5.21 Internal hardware of the four-channel Ames Spatial Auditory Display (ASAD). The prototype pictured here, fabricated for demonstration purposes, works in conjunction with KSC radio communication equipment. It includes an extra switch for a bypass mode that deactivates the spatial filtering and an extra knob for adjusting left and right line level test outputs (located on the back unit). The four digital boards (at top right and center) each contain a removable EPROM mounted on a ZIF socket, codec (Crystal CS4215), and DSP chip (Motorola 56001). The analog board (upper left) contains the input filtering and mixing circuitry. The power supply is at the bottom. *From Begault and Erbe (1993). Courtesy Audio Engineering Society.*

virtual auditory positions around the listener by digitally filtering each input channel with binaural HRTF data that are tailorable to the individual through the use of removable electrical-erasable-programmable-memory chips (EPROMs).

Figure 5.21 shows an overhead view of the ASAD and its various components. Input channels to the spatial auditory display can be assigned to any position by switching the EPROM used. The EPROMs contain binaural HRTF data for an arbitrary position and measurement source. While it is not necessary, different EPROMs can be installed or removed by a particular user who needs a special filtering set (such as ones based on personal HRTFs or to compensate for a particular set of headphones) by simply unplugging the old chip and installing the new one. One also can customize the ASAD's virtual positions in this way, although some applications would require arrangements known to optimize intelligibility.

The ASAD's interface was designed to be as uncomplicated as possible. Because each EPROM is associated with a particular HRTF and position, a user only needs to power the device up and adjust the overall volume. This integrated device approach is in contrast to spatial auditory displays that require a computer host, or a complex front panel interface for operation. Flexibility, while desirable in a recording studio, can be a nuisance or even a danger within a high-stress human interface context.

The ASAD's hardware is comprised of four DSP boards and an analog mixing/prefiltering board (see Figure 5.22). On each digital board is a DSP (Motorola DSP56001, running at 27 MHz), a codec (Crystal Semiconductor CS4215), three 32,768 x 8 static memory chips, and a 8,192 x 8 removable

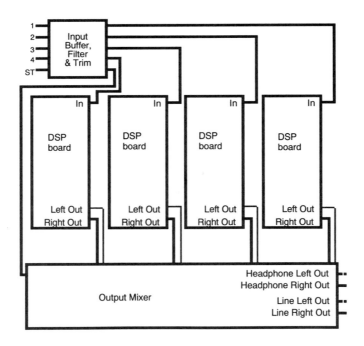

FIGURE 5.22. Schematic of the ASAD's internal circuitry. Each digital board functions to spatialize sound to a particular position (e.g., left 90 , left 30, right 30 and right 90 degrees azimuth). *Adapted from Begault and Erbe (1993). Courtesy Audio Engineering Society.*

EPROM. On the analog board are four low-pass filter modules, followed by four input gain trimpots. Also on the analog board are two five-channel mixers that matrix the left and right channel outputs. The analog board also contains trimpots for **sidetone** (the sound from the user's microphone) and headphone and line outputs. Sidetone is mixed equally to both channels without processing so that it appears in the center of the head, similar to the spatial audition of one's own voice.

The software for setting up the signal-processing chips specifies a time domain convolution algorithm for a mono-in—stereo-out FIR filter. There is also a provision for interchannel delay within the software (0–128 samples). The ASAD software is contained in the removable EPROM on each digital board, along with the filter coefficients and delay values. Each EPROM-DSP chip combination utilizes up to 224 coefficients (112 for each channel). This configuration posed a filter design problem for the available measured HRTFs that contained 512 coefficients per channel. The solution was to use synthetic HRTFs designed with FIR filter design algorithms, as described in Chapter 4.

An investigation into what single spatialized azimuth position yielded maximal intelligibility against noise was conducted using KSC four-letter call signs as a stimulus and speech babble as the masker (Begault, 1993a). This was accomplished by measuring thresholds for fifty-percent intelligibility at 30-degree azimuth increments, using a laboratory system that simulated the essential features of the ASAD. Figure 5.23 summarizes the data with the mean values for symmetrical left-right positions about the head. Not surprisingly, the most optimal positions for intelligibility is where interaural differences are maximized, in the vicinity of 90 degrees azimuth. A 6 dB advantage was found in the KSC call-sign experiment. The advantage found in a more controlled experiment using multiple male and female speakers (ANSI, 1989) was found to be 1 to 2 dB lower, except at 120 degrees azimuth where the advantage is about the same.

One unexpected advantage found with the ASAD is that it allows a more hands-free operation of the host communication device. Normally communication personnel bring the volume up for a desired channel on the front panel of the radio in order to hear that channel over undesired channels. With the ASAD, one can direct attention to the desired stream, as at a cocktail party. But another advantage is that overall intensity level at the headset can remain lower for an equivalent level of intelligibility, an important consideration in light of the stress and auditory fatigue that can occur when one is "on headset" for an eight-hour shift. Lower listening levels over headphones could possibly alleviate raising the intensity of one's own voice (known as the **Lombard Reflex**; see Junqua, 1993), reducing overall fatigue, and thereby enhancing both occupational and operational safety and efficiency of multiple-communication channel contexts.

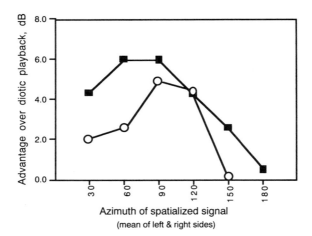

FIGURE 5.24. Speech intelligibility advantage of a 3-D auditory display compared to two-ear monaural (diotic) listening, as a function of virtual azimuths (mean of left and right sides). Mean values for fifty-percent intelligibility shown for five different subjects in each study. Black squares: Intelligibility advantage for spatialized KSC call signs against diotic speech babble, one male speaker (from Begault and Erbe, 1993). Circles: Intelligibility advantage using the Modified Rhyme Test (ANSI, 1989), using multiple speakers for the signal and speech-spectrum noise as the masker.

AERONAUTICAL APPLICATIONS

Head-up Spatial Auditory Displays

The current implementation of the Traffic Alert and Collision Avoidance System (TCAS II) uses both auditory and visual displays of information to supply flight crews with real-time information about proximate aircraft. However, the visual display is the only component delegated to convey spatial information about surrounding aircraft, while the auditory component is used as a redundant warning or, in the most critical scenarios, for issuing instructions for evasive action (Begault, 1993b).

Within its standard implementation, three categories of visual-aural alerts are activated by TCAS, contingent on an intruding aircraft's distance. The first category, an informational visual display, presents **proximate traffic.** In this case, TCAS functions more as a situational awareness system than as a warning system. The second category, a visual-aural cautionary alert, is a **traffic advisory.** The threshold for activating a traffic advisory is a potential conflict within forty seconds; an amber filled circle is generated on a visual map display, and an auditory warning consisting of a single cycle of the spoken words

"TRAFFIC, TRAFFIC" is given. The third category, a visual-aural warning alert, is a **resolution advisory.** The threshold for activating a resolution advisory is a potential conflict within twenty to twenty-five seconds; a red-filled square is generated on a visual map display, and an auditory warning enunciating the necessary appropriate evasive action (e.g., "CLIMB, CLIMB, CLIMB") is given.

Perrott *et al.* (1991) found that spatial auditory information can significantly reduce the acquisition time necessary to locate and identify a visual target. They used a 10 Hz click train from a speaker that was either spatially correlated or uncorrelated to a target light. The results showed that spatially correlated information from an auditory source substantially reduced visual search time (between 175–1200 msec depending on the target azimuth). In an experiment by Sorkin *et al.* (1989), localization accuracy rather than target acquisition time was studied in a simulated cockpit environment. A magnetic head tracker was either correlated or uncorrelated with a 3-D audio display that corresponded to the locations of visual targets. Results of the study found that accuracy of azimuthal localization was improved when head movement was correlated with the 3-D audio display but that elevation localization was no better than chance.

Begault (1993b) evaluated the effectiveness of a 3-D head-up auditory TCAS display by measuring target acquisition time. The experiment used pilots from a major air carrier flying a generic "glass cockpit" flight simulator. All crews used visual out-the-window search in response to a TCAS advisory, since no visual TCAS display was used. Half the crews heard the standard loudspeaker audio alert, and half heard an alert that was spatialized over headphones using 3-D sound techniques. The direction of the spatialization was linked to the target location's azimuth, but not its elevation. In addition, the spatialized audio stimuli were exaggerated by a factor of three in relationship to the visual angle to encourage head movement in the aurally guided visual search (e.g., visual targets at 10 degrees azimuth would correspond to spatialized stimuli at 30 degrees azimuth). Results of the study found a significant reduction in visual acquisition time by using spatialized sound to guide head direction (4.7 versus 2.5 sec).

A similar study by Begault and Pittman (1994) compared target acquisition time between a head-down visual display (standard TCAS) and a head-up audio display (3-D TCAS). 3-D sound was used for aurally guided visual search, as in the study by Begault (1993b), but without inclusion of the exaggeration factor mentioned above. Figure 5.24 shows the assignment of audio azimuths to visual positions and the available out-the-window field of view for the captain in the flight simulator (the first officer's view would be the mirror image of this figure). Flight simulators inherently have a limited field of view because of the use of screen displays for showing visual information relevant to the scenario. Note that the field of view from 25 to 52 degrees is available to only one crew

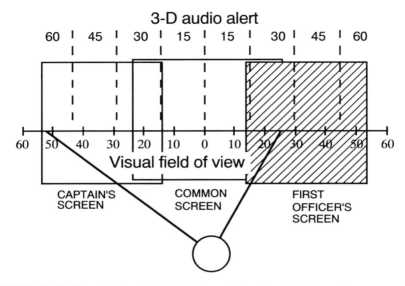

FIGURE 5.24. The horizontal field-of-view in the simulator from the perspective of the left seat (captain's position). The numbers within the dashed lines show the mapping between visual target azimuths and the specific azimuth position of the 3-D sound cue that was used for the TCAS alert. *Adapted from Begault and Pittman (1994).*

member, while the area between ±25 degrees is available to both crew members. The immediate range of the vertical field of view is from approximately –13 to +16 degrees but can extend from –18 to +20 degrees with head and body adjustments.

The mean acquisition time for the standard, head-down TCAS group was 2.63 sec, while the mean for the 3-D audio head-up group was 2.13 sec. There was no significant difference in the number of targets acquired between crews in each condition. The results imply that the presence of a spatial auditory cue can significantly reduce the time necessary for visual search in an aeronautical safety environment. This finding is in line with the studies of Perrott et al. (1991) that found advantages for aurally guided visual search using analogous conditions in the laboratory.

Although 500 msec may seem to be a modest improvement, it does suggest that, in an operational setting, an aural 3-D TCAS display may be desirable in addition to a standard TCAS display. This is because pilots can keep their head "out the window" looking for traffic without needing to move the head downward to the visual display and then back up. In other words, by accessing an alternative perceptual modality—sound—the visual perceptual modality is freed to concentrate on other tasks, if necessary.

AURALIZATION APPLICATIONS

The acoustical consultant is faced with many types of engineering challenges, including the control of intrusive or extrusive noise and vibration in rooms, and specification of materials and designs that can radically alter the acoustical and psychoacoustical impression of the environment. The noted acoustician Vern Knudsen wrote in 1963 in *Scientific American* that "Perhaps the biggest failure of architects (and acousticians) is in not doing something constructive about the acoustical environment of the home, particularly the apartment dwellings that have been built in the last twenty years" (Knudsen, 1963). He went on to make suggestions related to noise control that are taken up in many current building codes. However, his suggestion to cut the reverberation time of plaster-wall rooms in living spaces by at least half to improve quality of music and speech is still widely ignored. The loudest (most reflective) rooms in many homes are kitchens and children's bedrooms, which more often than not lack any sort of sound-absorbing material. Perhaps if these rooms had been "auralized" before their design, many of the living spaces in everyday life would be quieter, and more comfortable to live in. Unfortunately, in many contexts, acoustical design is often considered a luxury, a nuisance necessary for adhering to building noise codes, or a means for making sound sources louder and more intelligible.

Kleiner, Dalenbäck, and Svensson (1993) have given a list of auralization applications that go beyond the near-term domain of sound system evaluation. The list includes training (architects, acousticians, audio professionals, musicians, blind persons); factory noise prediction; studies in psychoacoustics related to enclosed environments; microphone placement evaluation; video game effects; automotive and in-flight sound system design; and cockpit applications.

The application of an auralization system to a given problem will result in accurate solutions only to the degree that both acoustical behavior and human perception are modeled accurately. For instance, the author found in one study that increasing the *number* of modeled early reflections influences the perceived spatial impression (envelopment) and distance of a sound source (Begault, 1987). In another study, using information derived from an acoustical CAD program, it was found that listeners could not discriminate between different spatial incidence patterns of six HRTF-filtered "virtual early reflections" (Begault, 1992c). Subjects auralized the convolution of test material under three configurations: The first was facing the sound source, as derived from a room model; the second, a version with the listener turned 180 degrees; and the third, with a *random* spatial distribution of reflections. The same timings and amplitudes were used for the early reflections in each case; only the directional information contained in the HRTFs was varied. Subjects were asked to state what differences they heard between the files, in terms of timbre, spatial positioning, or loudness of the source. They were also asked to state anything else they wanted to about the sound they heard.

Because most of the subjects used were expert listeners, all reported the sensation that the sound source was in some kind of room—i.e., they noticed that reverberation was present, although only early reflections were used. But none of the subjects reported the image switching back and forth from behind to in front of them while listening to the first and second patterns. In fact, it was extremely difficult for anyone to discriminate *any* difference between the three examples. The explanation given was that the spectral information in the HRTF filtered early reflections was masked by the direct sound and adjacent reflections (see Begault, 1992c). The use of a more complete model of early reflections might solve the problem, but it has not yet been determined how accurate a model needs to be. Another solution may be the inclusion of head-tracked virtual acoustics, as described by Foster, Wenzel, and Taylor (1991).

Progress continues to be made by several researchers in specific application areas of auralization. The Room Acoustics Group at the Chalmers University of Technology (Gothenburg, Sweden) published several papers evaluating the success of auralizing actual acoustic design projects (e.g., Dalenbäck, Kleiner, and Svensson, 1993). A particular area of concentration of the Bose® Corporation and its Modeler® group has been the verification of the prediction of speech intelligibility within an auralization system similar to that found in an actual room (e.g., Jacob, Jørgensen, and Ickler, 1992; Jørgensen, Ickler, and Jacob, 1993). They have proposed the term **authentication** for the process of quantifying "to what extent people hear the same thing in the simulated environment as they hear in the real environment." Mochimaru (1993) has studied methods for better modeling of on- and off-axis loudspeaker responses, and a research project initiated by RGP Diffusor Systems (Upper Marlboro, Maryland) hopes to gather measurements of directional scattering coefficients (DISC) for all commonly used architectural materials for auralization applications (D'Antonio, Konnert, and Kovitz, 1993). Many other basic research projects exist as well; the sum of all of these efforts, combined with further advances in DSP hardware, should make auralization a prime 3-D audio application area in the near future.

THE MUSIC OF THE FUTURE IS THE MUSIC OF THE PAST

If, paraphrasing Blauert (1983), it is an accurate assertion that there is no such thing as non-spatial hearing, then it logically follows that all musical experience has an inherent spatial component, regardless of the listener's attention. The inherent nature of aural communication requires a listener to occupy a different location from the sound source, and this is true of music if more than a single performer is involved. The compositional manipulation of the spatial aspect of music was as inevitable as the manipulation of pitch, timbre or rhythm, but little attention has been given to its long and varied historical development.

The ability of humans to localize sound is often cited as having evolved as an important mechanism for survival. When people or animals communicate across a physical distance, at least two types of information are transmitted: the semantic value of the sounds and the location of the sound sources. Early humans probably imitated animals for the purpose of establishing territory or for alerting others of danger. Birds communicate for this same purpose: "In the spring, male birds take possession of some patch of woodland, some lane, ledge, or meadow. When they do, they make every effort to keep out other males of the same species by a great deal of singing. . . . once the boundaries of a bird's territory have been set up, its song usually suffices to protect and preserve its property" (Theilcke, 1976).

This establishment of spatial relationships with sound by alternation between groups was taken early on into the context of religious ritual. The custom of **call and response** has its roots in the communication between tribal communities and in the interaction between the spoken words of a political or religious leader and the vocal response of their faithful. From a sociological perspective, the distance between a single person who speaks and a group of listeners establishes information about social rank (and sociological distance) between the two groups. All leaders must speak to those who follow them; the followers speak, shout, clap, or sing as part of the social dynamic of their response and to verify their relationship to one another and the leader through sonic means. The need to establish authority through ritual probably motivated the development of prescribed forms of single-person–multiple-person sonic interaction. (Television is a present-day example, although the authority of the sponsor is as much visual as it is sonic.)

Three techniques in organizing the spatial component of musical composition can be identified over the evolution of Western European art music (the type of music you learn about in music appreciation class): **Alternation** between locations, suggestion of **environments** and **distant sound sources**, and **sound source movement**. Figure 5.25 arranges these practices on a rough time line. While this categorization is arbitrary (and would constitute only the starting point for a thorough, musicological study), it does represent separable, identifiable trends in the historical development of spatial manipulation in music.

Alternation as a compositional technique involves developing a musical situation exploiting the exchange of phrases, strophes, or other musical units between two spatially separated sound sources or groups of sound sources. This technique culminated in the polychoral, or *cori spezzati* (separated choir), practice in Italy and Germany, which flourished around 1475–1640. Ensembles of vocalists and sometimes instrumentalists were divided into separate groups and the music would be performed either individually or together. The practice is often associated with St. Mark's Cathedral in Venice, where two organs were

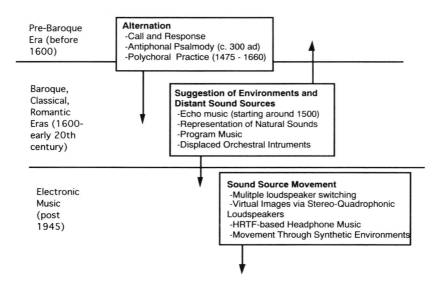

FIGURE 5.25. Time line describing the compositional practice of spatial manipulation in the Western art music tradition.

placed in opposite apses of the church. One polychoral mass, attributed to the composer Benevoli and written for the cathedral in Salzburg, was composed for fifty-eight voices in eight different groups; its musical interest "depends strongly on the element of surprise provided by the spatial separation" (Arnold, 1959) .

A curious variation on the technique of alternation is **echo music**, where music is composed specifically to imitate the echoes heard in natural environments. The device is found in music of the Renaissance, Baroque, and Classical eras (Laurie, 1980). An early example is the sixteenth-century madrigal by Orlando di Lasso, "O la, o che bon eccho." Later examples include the *Notturno en Echo* (K. 137) and the *Serenade* (K. 286) of Mozart and the aptly titled *Das Echo* by Joseph Haydn. The echoing instrument or voice is usually quieter, and always imitates the theme of the nonechoing sound source in close temporal succession.

Representation of natural phenomena by echo music is part of the larger trend of **musical representation of natural sounds**. Long before there were auditory icons representing everyday sounds, composers were notating music to suggest the sounds that would exist in a variety of environmental contexts. Examples from the era of Baroque music (around 1600–1750) include the use of cuckoo bird themes (usually a repeated descending minor third interval in even eighths), and the animal and various musical instrument sounds imitated by the solo violin in Carlo Farina's *Capriccio Stravagante* of 1627. By the time of Romantic-era operatic and orchestral composition, the use of sounds to suggest pastoral, battlefield, and other types of environmental contexts was commonplace.

The manipulation of **auditory distance** was brought about particularly in Romantic-era orchestral music, where timbral and dynamic changes were easily facilitated. Manipulation of auditory distance in instrumental music can be produced by varying dynamic indications or by reducing the number of instruments in an orchestral desk. Alternatively, the illusion of sounds in the distance was effected by amplitude-timbral changes written into the orchestration of particular instruments, e.g., the use of **mutes** on brass instruments. A more complex approach was used by those composers who placed instruments behind a curtain on the side of a stage to effectively reduce the high-frequency content of a source as well as its amplitude. Sound sources at different distances give rise to a sense of the **environmental context;** orchestral music created aural virtual realities not only through distant, imitative sounds, but also by writing **program music.** For example, one doesn't realize until they've read the program that one is hearing a "March to the Scaffold" in Hector Berlioz's *Symphonie Fantastique*. This common practice continues to this day: a composer (or seller of a 3-D audio product) supplies the listener with a verbal or written description of what they're supposed to be hearing.

Spatial movement was first effected through the alternation of signals between loudspeakers. Later it developed into a practice of creating the illusion of auditory movement. The advent of loudspeakers and electrical control devices offered the possibility of a different approach to spatial manipulation. For distance, the musical techniques involving the literal separation of sources were understood early on, but the later trend of controlling auditory distance from instruments in a standard orchestral arrangement—i.e., the use of virtual spatial imagery—has extended into musical composition that uses the tools of modern recording technology.

An early example of an electronic music composer who methodically explored the compositional control of distance was Karlheinz Stockhausen. His *Gesang der Jünglinge* (1955) used the technique of altering the proportion of reverberant and direct sound and intensity of the spoken voice for creating distance effects. In contrast, there are composers who manipulated sound space by locating multiple speakers at various locations in a performance space and then switching or panning the sound between the sources. In this approach, the composition of spatial manipulation is dependent on the location of the speakers and usually exploits the acoustical properties of the enclosure. Examples include Varese's *Poeme Electronique* (tape music performed in the Philips Pavilion of the 1958 World Fair, Brussels) and Stanley Schaff's *Audium* installation, currently active in San Francisco.

In the 1970s, quadraphonic playback was the rage for a brief period in commercial music, and for a longer time among electronic and computer music composers. Several composers, among them Maggi Payne, Roger Reynolds, and John Chowning, wrote electronic music specifically for this domain that manipulated the possibility of sound trajectories—patterns of sound movement that included the virtual space between the loudspeakers. As one might expect, the concert playback experience never matches the experience of listening in the sweet spot of the composer's recording studio in terms of the perception of sound movement in this music (although this is irrelevant to enjoyment of the music's unpredictable spatial qualities).

One of the first major spatial manipulation programs for audio came out of quadraphonic loudspeaker playback techniques for computer music. Chowning's *Simulation of Moving Sound Sources* (1971) specified an algorithmic control method for both sound movement and distance and was used for many computer music compositions. The reverberant sound was distributed in two ways: either to all four speakers or to an individual speaker along with the direct sound. These techniques were termed **global reverberation** and **local reverberation**, respectively. The motivation for this design was to simulate the change in the spatial aspects of reverberation with distance: As a sound moved in closer, global reverberation from all four speakers increased; and as the sound moved out further from the modeled listener position, local reverberation increased at the particular loudspeaker. This approach is essentially a modeling of the change in auditory spaciousness with distance. A variant of this program that found use in many applications was described in Moore (1983). This program, a part of the **cmusic** software synthesis package, featured ray tracing for early reflections in a two-dimensional room model. An "inner room" was modeled for the listener, with "windows" looking out into an "outer room" occupied by the sound source; rays were traced from the source to the window openings, which could represent loudspeakers or headphone drivers (Moore, 1990).

3-D Techniques Applied to Headphone Music

3-D audio technology was used by the author to create a headphone music composition entitled *Revelations*, where sound movement patterns were a key feature (Begault, 1987). This composition uses both spatial manipulation and juxtaposition of text fragments as the principal type of compositional grammar. The work was realized with a software synthesis program (cmusic). The text used in the composition, an extract of which is shown in Figure 5.26, is based in part on a segment from a poem by John Giorno. Sampled sounds, consisting of plucked piano strings and violin special effects, were used to accompany a spoken narration of the poem. The material was recorded into the computer in small segments, and then multiple copies were made by HRTF filtering the sound to twelve positions on the azimuth at 0 degrees elevation. The software synthesis program was used to fade between the sound files at various amplitudes so as to create smooth spatial trajectories.

```
you are bored
you are bored
you are bored       and restless
you can't think of anything to do
to do
to do
to do
to do
you can't think of anything to do

maybe it will get better if you walk around

so you walk around
and around
and around
and around
and around
```

FIGURE 5.26. Two excerpts from "We Got Here Yesterday, We're Here Now, and I Can't Wait to Leave Tomorrow" by John Giorno (1981). The boldface portions are used to show the spatialized parts of the text and were not boldfaced in the original poem. *From* Grasping at Emptiness *(1985), published by Kulchur Foundation, New York, New York. Used by permission.*

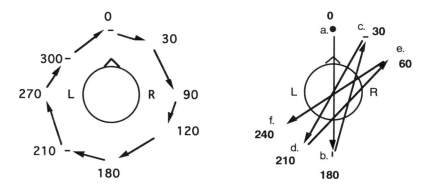

FIGURE 5.27. Spatial trajectories for the phrase *to do* used in the computer music 3-D sound headphone composition *Revelations* (1987). On the left side, an erratic trajectory moves between each point across the same period of time, but alternates between 30- and 60-degree jumps. On the right side, a see-saw trajectory has been set up. *Adapted from Begault (1987).*

The entire reading of the text was delivered diotically, resulting in an inside-the-head spatial position for the narration—except for the sound files containing the phrases "to do" and "and around." These phrases were spatialized in the same manner as the instrumental sounds so that they jumped out of the overall spatial texture and literally outside of the head along with the accompanying instrumental sounds.

There were several types of spatial sound movement, or **trajectories,** that were composed in advance and could be applied to any of the spatialized sounds (see Figure 5.27). A spatial aspect that remained constant for all trajectories was "circularity," defined here as moving once completely around the listener, by whatever means. The spatial aspects that were variable included the kind of pattern used to attain circularity, the overall time required to complete the pattern, and the direction of circularity (clockwise or counterclockwise). In this way, a compositional approach to spatial movement could be made analogous to melodic "theme" and "variation": One could refer to "melodies of spatial movement."

The basic spatial theme for this example was an eight-second counterclockwise circle used for the piano and violin sounds. To form this trajectory, a series of twelve attacks at successive 30-degree positions was created, with about half a second between each attack. The piano and violin sounds each had a duration of about 1.5 seconds; as a result, the decay of each previous attack overlapped with the attack of the next sound. The perceived result was of sound movement, rather than of a sound alternating between twelve discrete points. The violin and piano were spatially separated by 180 degrees before beginning the trajectory, so that one sound followed the other at the opposite position on the azimuth. When listening to these two 180-degree opposed, circular-moving sounds, there seemed to be two ways to focus attention: either to hear the two streams simultaneously as a unit, or to focus one stream only. In the case of listening to both streams, one has the novel impression of being at the pivot of an axis with a sound source at either end.

Two other types of trajectories were used as variants, illustrated in Figure 5.27. The first trajectory shifted positions in 0.5 sec. intervals, moving clockwise 30 degrees and then 60 degrees around the listener. This trajectory was meant to be a variation on the circular trajectory discussed above; it differs in the use of an opposite (clockwise) direction and use of faster, erratic movement caused by the 30- to 60-degree leaps. Also, the movement effect discussed previously did not result because the sound was around 0.5 seconds in duration, with no overlap between attacks. The second trajectory shifted positions in 0.25 sec intervals to the 180-degree opposite position and then back 150 degrees: 0 –180–30–210–60–240, until all twelve positions were articulated (right side of Figure 5.27). This procedure resulted in a rapid see-saw effect, chosen to simultaneously alternate from side to side while maintaining the circular identity of the other trajectories.

Gordon Mumma: Audioearotica

A compositional sketch produced for headphone listening by Gordon Mumma in conjunction with the author probably represents one of the first combinations of 3-D sound and auralization techniques for a headphone music composition. Subtitled *Begault Meadow*, it uses only a single, short sound: a 50-msec pulse of relatively broad band noise with a mildly tapered attack and decay. The landscape of the simulated meadow was differentiated by modeling a grassy meadow to the rear, a steeply rising hill to the right-front, and a lake to the left-front. The simulation was achieved through specification of frequency-dependent absorption for each modeled wall' (direction) specified to a customized auralization program written by the author.

Mumma wrote that "The musical syntax is achieved entirely with the spatial placements of that single sound; there are no pitch, timbre, or loudness attributes except as they result from spatial location of the single sound. The poetic impetus for the music derives from Abbott's classic book *Flatland;* to support the 'meadow' part of the title, the acoustical characteristics of an outdoor environment are synthesized. . . . The spatial choreography of *Begault Meadow* varies from one section to another of the piece. In some sections the sounds clearly inhabit specific azimuth regions around the listener; in other sections the location of sounds is more ambiguous. The play between specificity and ambiguity is largely a function of compositional choices" (Mumma, 1986).

Because of the implementation of significant physical parameters of a modeled outdoor environment into the auralization program used to create the sounds, the noise burst's timbre changed significantly as a function of direction and distance. One consequently has a more realistic sense of spatial extent and of being surrounded by the sound source, since the environment is differentiated, as in the real world. Since few of us have experience of a 50-msec burst of white noise in a meadow, one does not hear the work and immediately recall such an environment. If one wanted to create such associations for the listener, it would be best to use cognitive cues: For example, a cow mooing, wind through the grass, and insect sounds. The meadow of this work is a virtual, arbitrary creation that from one perspective requires learning through repeated experience (listening).

Audioearotica and *Revelations* are but two examples of the potential application of the HRTF signal-processing technique to the composition of spatial music. Headphone music continues to be a largely unexplored territory of the infrequently visited universe of spatial music, and we should expect our "mind's aural eye" to be further surprised and challenged in the future.

CHAPTER SIX

Resources

FINDING PERTINENT INFORMATION

O nce committed, a person probably can never have enough information on a topic as broad as sound and its implementation within the emerging virtual worlds of entertainment, communication and business. Unfortunately, there never seems to be enough time to get through all of the information, and false leads often lead to an abundance of time wasted looking for a particular nugget of information that will make an application fly like an eagle instead of a dodo. Furthermore, the rapid turnaround cycle for audio products makes recurrent inventories of available technologies from major manufacturers important.

Some of the topics covered in this book will require a more in-depth treatment. The perspective for designing a successful virtual acoustic display offered in this book involves disciplines as divergent as psychoacoustics, electrical engineering, and music; more than likely, you're very familiar with some of these areas and new to others. Each topic requires elucidation, not only separately but especially in terms of the overall connection. Because of this, an attempt has been made throughout this book to keep technical explanations on an accessible level, so the reader would not lose sight of the "big picture," the inevitable link between the listener's spatial hearing experience and the technologies used to bring about 3-D sound.

To help the reader get started, this chapter contains the names and addressees of several useful journals, books, magazines, and companies that are particularly pertinent to 3-D sound. This list is by no means all-inclusive; for instance, information on many "loudspeaker only" 3-D audio products, commercial reverberators, and surround-sound processors will not be found here. These resources are offered to give one access to more in-depth tutorial-level explanations; advanced, technically oriented texts; journals and proceedings for more up to the moment information; and hardware and software tools for manipulating sound.

BOOKS AND REVIEW ARTICLES

Virtual Reality

Inspirational and Informative Works

Jacobson, L. (Ed.) (1992). **Cyberarts. Exploring Art and Technology.** San Francisco: Miller Freeman.

> *An easy-to-read overview of technologies relevant to many types of new interactive media, including tutorial lectures on 3-D sound by Linda Jacobson, Durand Begault, Elizabeth Wenzel, Curtis Chan, Scott Foster, and Christopher Currell.*

Rheingold, H. (1991). **Virtual Reality.** New York: Summit Books.

Krueger, M. W. (1991). **Artificial Reality II.** Reading, MA: Addison-Wesley.

Laurel, B. (1991). **Computers as Theater.** Reading, MA: Addison-Wesley.

> *The books by Rheingold, Krueger, and Laurel have inspired many workers and thinkers in the fields of virtual reality, and are required reading for all interested parties.*

Light Technical Overviews

> *Since about 1991, there has been an explosion of books on the topic of multimedia and virtual reality, aimed at different audiences and ranging widely in quality. Two of the better ones are:*

Pimentel, K., and Teixeira, K. (1993). **Virtual Reality. Through the New Looking Glass**. Blue Ridge Summit, PA: Windcrest.

Wexelblat, A. (Ed.). (1993). **Virtual Reality. Applications and Explorations.** Cambridge, MA: Academic Press.

Heavy-weight Technical Overviews

> *While the "gee-whiz" factor is featured in many virtual reality books, the following titles are representative of more technically-oriented treatments.*

Kalawsky, R. S. (1993). **The Science of Virtual Reality and Virtual Environments.** Wokingham, England: Addison-Wesley.

One of the better overviews of virtual reality systems. The book accounts for human factors and perceptual issues and includes more technical information for systems design than found in most books on virtual reality.

Ellis, S. R., Kaiser, M. K., and Grunwald, A. J. (Ed.) (1993). **Pictorial Communication in Virtual and Real Environments.** London: Taylor and Francis.

While devoted to visual rather than auditory topics, this book represents one of the few efforts to integrate the perceptual, physiological, and engineering issues related to visual displays, particularly virtual reality.

Multimedia

Blattner, M., and Dannenberg, R. (Ed.) (1992). **Multimedia Interface Design.** Reading, MA: ACM Press/Addison-Wesley.

Buford, J. (Ed.) (1994). **Multimedia systems.** Reading, MA: ACM Press/Addison-Wesley.

Multimedia is an area in which there is a glut of "how-to" books that generally focus on uninteresting applications of sound. These two books are exceptions. The book edited by Blattner and Dannenberg is highly theoretical while the papers in Buford's book are more technically oriented.

Sound and Psychoacoustics

Pierce, J. R. (1983). **The Science of Musical Sound.** New York: Scientific American Books.

An easy-to-read, nicely illustrated book useful for understanding the basic physical and psychoacoustical issues of acoustics; although the primary subject is music, it's really about sound of all types. Probably one of the most accessible books for a beginner.

Handel, S. (1989). **Listening: An Introduction to the Perception of Auditory Events.** Cambridge, MA: MIT Press

Another good first book for understanding the relationship between the physics and psychophysics of sound.

Moore, Brian C. J. (1989) **An Introduction to the Psychology of Hearing** (3rd ed.). New York: Academic Press.
A college-level introductory review of issues relevant to psychoacoustics, including a good chapter on space perception.

Roederer, J. G. (1979). **Introduction to the Physics and Psychophysics of Music.** (2nd ed.). New York: Springer-Verlag.
More advanced than Pierce's Science of Musical Sound, *this college-level introductory text covers hearing, physics, psychophysics, and neuropsychology.*

Digital Audio and Computer Music

Dodge, C., and Jerse, T. A. (1985). **Computer Music: Synthesis, Composition, and Performance.** New York: Schirmer Books.

Moore, F. R. (1990). **Elements of Computer Music.** Englewood Cliffs, NJ: Prentice-Hall.
This book not only covers most topics related to audio DSP, it also has C programming examples for do-it-yourself implementation and an entire chapter devoted to "Rooms"— including algorithms for sound spatialization. Extremely useful reference for digital synthesis and filtering techniques, as well as mathematics.

Pohlmann, K. C. (1992). **The Compact Disc Handbook** (2nd ed.). Madison, WI: A-R Editions.

An excellent writer, Pohlmann demystifies sampling rates, quantization, and how a CD player really works. The Compact Disc Handbook *is an especially good first introduction to digital sound.*

Oppenheim, A. V., and Schafer, R. W. (1975). **Digital Signal Processing.** Englewood Cliffs, NJ: Prentice-Hall.

Oppenheim, A. V., and Schafer, R. W. (1989). **Discrete-time Signal Processing.** Englewood Cliffs, NJ: Prentice-Hall.

These are the classic technical references for digital filter theory, used as college texts in many signal-processing courses.

Moore, F. R., "An Introduction to the Mathematics of Digital Signal Processing"; Smith, J. O., "Introduction to Digital Filter Theory"; Moorer, J. A., "Signal Processing Aspects of Computer Music: A Survey" *in* Strawn, J. (Ed.) (1985). **Digital Audio Signal Processing: An Anthology.** Madison, WI: A-R Editions (formerly published by W. Kaufmann, Los Altos, CA).

> *For those first delving into the thorny bushes of digital signal processing and digital filters, these papers and other relevant articles in this collection are essential reading.*

Sound System Design

Davis, G., and Jones, R. (1989). **Sound Reinforcement Handbook** (2nd ed.). Milwaukee, WI: Hal Leonard Publishing.

Runstein, R. E., and Huber, D. M. (1989). **Modern Recording Techniques** (3rd ed.). Indianapolis, IN: Howard W. Sams.

> *These two titles are invaluable references for audio applications and include introductory explanations of topics related to sound recording and signal processing.*

Rothstein, J. (1992) **MIDI: A Comprehensive Introduction.** Madison, WI: A-R Editions.

> *There seem to be hundreds of books on MIDI available on the market today. This book (as well as the several written by Steve De Furia and Joe Scacciaferro) is a good choice but inspect any book before you buy it—one may be more appropriate than another for your particular needs. MIDI books that feature interviews with musicians about their particular implementation are probably the least useful of the genre. Many of the books reprint a copy of the actual MIDI specification, which is available separately:*

MIDI Manufacturers Association Technical Standards Board. **MIDI 1.0 Detailed Specification.** Version 4.2. Los Angeles: International MIDI Association.

> *A must-have reference for programming applications. Available from the International MIDI Association, 11857 Hartsook Street, North Hollywood, CA 91607.*

Spatial Hearing References

Gatehouse, R. W. (Ed.) (1982). **Localization of Sound: Theory and Applications.** Groton, CT: Amphora Press.
Collected papers from a conference on spatial hearing. Includes important papers from E. Shaw (on HRTF measurement), J. Blauert (multiple images and applications in room and electroacoustics), and N. Sakamoto (an early paper on localization of virtual acoustic images).

Blauert, J. (1983). **Spatial Hearing: The Psychophysics of Human Sound Localization** (translation of **Räumliches Hören**, J. Allen, trans.). Cambridge, MA: MIT Press.
This is the classic reference book in the field of auditory spatial perception; it's hard to imagine anyone working in 3-D sound not having referred to it at one time or another. Includes references to many European studies unavailable in English and an excellent bibliography.

Yost, W. and Gourevitch, G. (Eds.) (1987). **Directional Hearing.** New York: Springer Verlag.
A collection of scholarly papers concerned with the physiology and psychoacoustics of spatial hearing.

Middlebrooks, J. C., and Green, D. M. (1991). **Sound Localization by Human Listeners.** *Annual Review of Psychology* **42**:135-159.
A compact overview of the main issues related to the psychophysics of spatial hearing.

Room Acoustics and Perception

Cremer, L., and Muller, H. A. (1982). **Principles and Applications of Room Acoustics** (Translation of **Die wissenschaftlichen Grundlagen der Raumakustik**, T. J. Schultz, translator). London: Applied Science.
A two-volume set comprising an extensive overview of the physics and psychoacoustics of rooms. A basic technical reference for those delving into auralization system design.

Ando, Y. (1985). **Concert Hall Acoustics.** Berlin: Springer-Verlag.
A compact book that describes in detail a case study method for calculating the psychoacoustic preference of an arbitrary concert hall design. A very relevant book for auralization system implementation.

Beranek, L. L. (1992). **Concert Hall Acoustics-1992.** *Journal of the Acoustical Society of America* **92(1)**:1-39.
An excellent tutorial and review on room acoustics and psychoacoustic issues related to reverberation. *Beranek's classic* Music, Acoustics and Architecture *(published in 1962) is currently out-of-print but available at many libraries.*

ANTHOLOGIES AND PROCEEDINGS

Audio Engineering Society (1993). **The Sound of Audio.** **Proceedings of the AES 8th International Conference**

Audio Engineering Society (1993). **The Perception of Reproduced Sound.** **Proceedings of the AES 12th International Conference**
These two volumes consist of papers from conferences where perceptual issues of sound reproduction, including 3-D sound, were featured topics.

Audio Engineering Society (1986). **Stereophonic Techniques.**
An anthology of classic articles on the reproduction of stereo sound, including many hard-to get articles from the 1950s.

Kleiner, M. (Ed.) (1993). *Journal of the Audio Engineering Society* **41(11)**, special issue on auralization.
Several articles on auralization techniques, rewritten and edited from AES conference preprints from 1992. Kleiner's introduction includes a useful bibliography.

All four of these titles are available from:
 Audio Engineering Society
 60 East 42nd Street
 New York, NY 10165-2520

Kramer, G. (Ed.) (1994). **Auditory Display. Sonification, Audification, and Auditory Interfaces.** Reading, MA: Addison-Wesley.
Collected papers of the proceedings of the First International Conference on Auditory Display, held in 1992 in Santa Fe. An important book on the uses of nonspeech audio, including several papers that consider the use of 3-D auditory displays. An audio CD is included that includes demonstrations of the speech intelligibility and aeronautical applications of 3-D sound discussed in this book and many other interesting examples.

Proceedings of the International Computer Music Conference. San Francisco: International Computer Music Association.

Since 1974, the International Computer Music Conference has gathered workers in various fields of music, acoustics, and computers. Scholarly articles and highly experimental approaches to music composition, real-time performance sensing, MIDI, etc. Available from the International Computer Music Association, P. O. Box 1634, San Francisco, CA 94101.

Proceedings of the Association for Computing Machinery (ACM) Special Interest Group on Computer and Human Interaction (SIGCHI). Reading, MA: ACM Press/Addison-Wesley.

These proceedings are of particular interest to those involved in designing auditory displays.

JOURNALS

Computer Music Journal
MIT Press Journals
55 Hayward Street
Cambridge, MA 02142-9902

Scholarly articles and technical reviews of computer music software synthesis techniques, compositional theory and algorithms, new interactive devices, etc. The standard resource for new developments in computer music since the late 1970s.

Electronic Musician
Mix Magazine
P. O. Box 41525
Nashville, TN 37204

These two magazines are aimed at professional music and recording audiences and cover a wide range of topics. Both magazines have featured several interesting articles on signal processing and 3-D sound, written at a nontechnical level.

Human Factors: The Journal of the Human Factors and Ergonomics Society
P. O. Box 1369
Santa Monica, CA 90406-1369

As 3-D sound moves toward applications in aeronautics and other fields, more specialized journals such as this have contained a number of research papers related to 3-D sound.

Journal of the Acoustical Society of America
500 Sunnyside Boulevard
Woodbury, NY 11797

A basic research journal containing articles on all aspects of acoustics. Most scholarly papers devoted to basic research on sound localization have appeared in this journal.

Journal of the Audio Engineering Society
60 East 42nd Street
New York, NY 10165-2520

More oriented toward audio professionals and applications, this publication has featured a number of important articles on sound localization and 3-D audio systems.

Presence: Teleoperators and Virtual Environments
MIT Press Journals
55 Hayward Street
Cambridge, MA 02142-9902

Several articles related to 3-D sound have appeared in this relatively new journal.

HRTF MEASUREMENTS

Shaw, E. A. G., and Vaillancourt, M. M. (1985). **Transformation of sound-pressure level from the free field to the eardrum presented in numerical form**. *Journal of the Acoustical Society of America* **78**:1120-1122.

This is an excellent source for averaged HRTF magnitude transfer functions that are based on the article by Shaw (1974).

Recently a set of KEMAR HRTF measurements have been available on the Internet by Bill Gardner and Keith Martin of MIT's Media Lab. Here is a condensation of the information they have provided:

The measurements consist of the left and right ear impulse responses from a loudspeaker mounted 1.4 meters from the KEMAR. Maximum length (ML) pseudo-random binary sequences were used to obtain the impulse responses at a sampling rate of 44.1 kHz. A total of 710 different positions were sampled at elevations from -40 degrees to +90 degrees. Also measured were the impulse response of the speaker in free-field and several headphones placed on the KEMAR. This data is being made available to the research community on the Internet via anonymous ftp and the World Wide Web.

The data is organized into two tar archives, accompanied by a document (Postscript and text) describing the measurement technique and data format. To access the files via ftp, connect to `sound.media.mit.edu`, login as "anonymous", and give your Internet address as a password. The files may be found in the directory "pub/Data/KEMAR". There is a README file describing the contents of this directory.

To retrieve the HRTF data via the World Wide Web, use your browser to open the following URL:

`http://sound.media.mit.edu/KEMAR.html.`

3-D SOUND SYSTEMS AND AUDIO CARDS

The main focus of companies that manufacture sound cards that plug into computer bus slots is games, since these sell millions of units. Closely related are the multimedia cards that contain a simple FM or wave table synthesizer that can be linked to visual presentation software. By contrast, professional-quality sound cards and related software intended for acoustic analysis or professional sound editing and digital signal-processing are few. No attempt has been made to catalogue all of the currently available sound cards; a visit to the local computer store will bring the interested reader up to date more effectively. Examples of some of the hardware and software particularly amenable to 3-D sound are listed, by virtue of (1) audio quality, (2) sophistication of the interface, and/or (3) availability of third-party software development support.

3-D Audio Systems

Crystal River Engineering
490 California Avenue, Suite 200
Palo Alto, CA 94306

> *Manufacturers of the Convolvotron, Beachtron, and Acoustetron 3-D audio devices described in detail within this book. They plan to offer a personal HRTF measurement system in the near future.*

Focal Point
9 Downing Road
Hanover, NH 03755

> *Producers of plug-in 3-D audio cards for Macintosh and PC platforms, described in Chapter 5.*

Roland Corporation (USA)
7200 Dominion Circle
Los Angeles, CA 90040-3696
Suppliers of products utilizing Roland Sound Space Technology (RSS) described in Chapter 5.

HEAD acoustics GmbH
Kaiserstrasse 100
D-5120 Herzogenrath 3 Germany
Distributed in North America by
Sonic Perceptions, Inc.
28 Knight Street
Norwalk, CT 06851
Manufactures the BMC (Binaural Mixing Console) and other integrated 3-D audio products (see "Headphones and Dummy Heads" section later in this chapter).

Sound Systems (Synthesis, Analysis, and Hardware)

Advanced Gravis Computer Technology Ltd.
101 North Fraser Way
Burnaby, British Columbia
Canada V5J 5E9
Multimedia sound card primarily for games, notable in its inclusion of Focal Point Technology for sound spatialization.

Ariel Corporation
433 River Road
Highland Park, NJ 08904
Audio signal-processing cards for a wide range of platforms and software.

Digidesign
1360 Willow Road, Suite 101
Menlo Park, CA 94025
Manufactures the Motorola 56001-based AUDIOMEDIA II™ multimedia card and SoundTools™ and ProTools™ professional audio systems. Good support for third-party developers. The Sound Designer™ software package is one of the more popular digital editing systems.

Oros Signal Processing
13 Chemin de Prés BP 26 ZIRST
38241 Meylan
France

Manufactures the OROS-AU-22 system, an audio signal-processing board utilizing a Texas Instruments TMS320-C25 as the primary DSP chip. Used for HRTF measurement by the room acoustics group at the Institute for Research Coordination in Music and Acoustics (IRCAM), Paris, France.

Sonic Solutions
1891 East Francisco Boulevard
San Rafael, CA 94901

Manufactures high-end, professional digital audio editing hardware and software for Macintosh platforms, including CD mastering and networking systems. Sonic Solutions also manufactures a system for removing noise and unwanted artifacts from recordings, called NoNOISE.®

Symbolic Sound Corporation
P. O. Box 2530
Champaign, IL 61825-2530

Manufactures Kyma™ software and Capybara™ hardware systems for software synthesis using Macintosh platforms. Very professional-level material.

Turtle Beach Systems
P. O. Box 5074
York, PA 17405

Manufactures the MultiSound™ multimedia sound card for PC compatibles and a number of other software and hardware products for various platforms.

HEAD TRACKERS AND OTHER INTERACTIVE SENSORS

The following lists three of the more popular head-tracking system manufacturers, along with the alternative boom-mounted display made by Fakespace, Inc.. The head-tracking devices represent the "middle range" of accuracy and/or expense.

Ascension Technology Corporation
Attn. Jack Scully
P. O. Box 527
Burlington, VT 05402

Manufactures the Bird and A Flock of Birds magnetic sensors.

Fakespace, Inc.
4085 Campbell Avenue
Menlo Park, CA 94025

Manufactures the counterbalanced BOOM (binocular omni-orientation monitor) for interactive virtual experiences without goggles.

Logitech
6505 Kaiser Drive
Fremont, CA 94555
Manufacturers a line-of-sight ultrasonic head-tracker and the 2D/3D Mouse.

Polhemus
P. O. Box 560
Colchester, VT 05446
Manufactures the Isotrak and Fastrak six degree-of-freedom magnetic sensor systems. The InsideTRAK allows the peripheral hardware to insert directly into a PC-compatible bus slot.

Auralization Tools

CATT-Acoustic (Computer-Aided Theatre Technique)
Attn. Bengt-Inge Dalenbäck
Svanebäcksgatan 9B
S-414 52 Gothenburg
Sweden
A room acoustic prediction and auralization program using the Lake DSP-1 convolution device. Represents the research of Mendel Kleiner (Chalmers Room Acoustics Group, Department of Applied Acoustics, Chalmers University of Technology, Gothenburg, Sweden), Peter Svensson, and Bengt-Inge Dalenbäck.

EASE-EARS
ADA (Acoustic Design Ahnert)
Berlin D-0 1086
Germany

Renkus-Heinz, Inc. (English language version distributor)
17191 Armstrong Avenue
Irvine, California 92714
Distributes the EASE and EASE Jr. room design programs for PC compatibles and the EARS auralization system (based on several plug-in card configurations for cards manufactured by Media Vision and Ariel). Represents the work of Dr. Wolfgang Ahnert and associates at ADA. Dr. Ahnert's papers on his implementation technique are also worthwhile: e.g., Ahnert and Feistel (1991).

Lake DSP, Pty. Ltd.
Suite 4, 166 Maroubra Road
Maroubra 2035
Australia
  Manufacturers of the FDP-1 series of large-scale convolution devices.

MODELER®
Sound System® Software
Bose® Corporation
Pro Products, Mail Stop 233
The Reservoir
Framingham, MA 01701
  MODELER is a room design program for Macintosh platforms. It includes speech intelligibility prediction algorithms (but an auralization system is not yet available commercially). Represents the research of Kenneth Jacob, Chris Ickler, and Tom Birkle.

HEADPHONES AND DUMMY HEADS

AKG (Akustische und Kino-Geräte GmbH)
Brunhildengasse 1, P.O.B. 584
A-1150 Vienna, Austria
AKG Acoustics Inc. (USA)
1525 Alvarado St.
San Leandro, CA 94577
  As of this writing, the BAP-1000 3-D audio device (discussed in Chapter 5) is only available in Europe.

Brüel and Kjaer
DK-2850 Naerum
Denmark
  Brüel and Kjaer offer a wide line of analysis equipment, microphones, and dummy heads relevant to 3-D audio production. Their representatives are available throughout the world.

Etymotic Research
61 Martin Lane
Elk Grove Village, IL 60007
  Manufactures insert earphones, including the flat frequency response ER-2 Tubephone™, hearing protectors, and other audiological tools.

HEAD acoustics GmbH
Kaiserstrasse 100
D-5120 Herzogenrath 3
Germany
Distributed in North America by
Sonic Perceptions, Inc.
28 Knight Street
Norwalk, CT 06851
> *Provides an extensive line of analysis equipment, dummy heads, and playback systems related to spatial audio production, noise control, binaural mixing consoles, etc.*

Knowles Electronics, Inc.
1151 Maplewood Drive
Itasca, IL 601433
> *Manufacturers of the famous KEMAR mannequin and torso.*

Georg Neumann GmbH
Charlottstrasse 3 D-1000
Berlin 61
Germany
Neumann USA
4116 West Magnolia, Suite 100
Burbank, CA 91505
> *Manufactures the KU-100 dummy head and high-end professional audio microphones.*

Sennheiser Electronic K.G.
Am Labor 1
30900 Wedemark, Germany
Sennheiser Electronic Corporation (USA)
6 Vista Drive
Old Lyme, CT 06371
> *Manufacturers of headphones, headsets, microphones, noise cancellation technology; MKE-2002 Binaural Head*

RELEVANT SOFTWARE

Computer Audio Research Laboratory (CARL) Software Distribution (including cmusic)
University of California, San Diego
La Jolla, CA 92093

An excellent audio software synthesis package for the UNIX environment, developed primarily during the 1980s at the Computer Audio Research Laboratory (University of California, San Diego). It includes F. R. Moore's cmusic and many interesting programs primarily by Moore, G. Loy, and M. Dolson. An upcoming version of cmusic, pcmusic, will be run under the Windows™ environment. Available via anonymous ftp from `ccrma-ftp.stanford.edu.`

Csound

MIT Media Laboratory
20 Ames Street
Cambridge, MA 02139

Software synthesis for a wide variety of platforms. Available via anonymous ftp from `ems.media.mit.edu.`

Opcode Systems

3950 Fabian Way, Suite 100
Palo Alto, CA 94303

MAX™ is a graphical real-time object-oriented programming environment for Macintosh platforms. It is especially useful for interfacing real-time control between MIDI-based sensors and effectors; it works also with CD-ROM drives, laser discs, and multimedia software. Opcode Systems also offers a wide range of MIDI interfaces and software.

SoundHack

Written by Tom Erbe
Frog Peak Music
P.O. Box 5036
Hanover, NH 03755

This Macintosh program has been described as a "Swiss Army knife" for software synthesis, since it contains many useful utilities for changing the spectral and temporal features of a sound file with optimized DSP programs. It includes a simple HRTF-filtering package.

Zola Technologies

6195 Heards Creek Dr., N.W.
Suite 201
Atlanta, GA 30328

For Macintosh platforms, Zola's DSP Designer provides a powerful environment for development of digital signal-processing algorithms and software. It is actually a set of tools and scripts that expand the basic Apple MPW environment to allow design, analysis, and floating-point simulation of DSP systems, including tools for creating assembly code for the

Motorola 56001. Record and playback capabilities are enabled via the Digidesign series of DSP cards, e.g., AUDIOMEDIA II.

3-D Recordings

The Binaural Source
P.O. Box 1727
Ross, CA 94957
To obtain "the world's only catalog of exclusive recordings for Headphone Experiences," write to the Binaural Source for their free catalogue. This is a small company run by John Sunier, a writer who has advocated binaural recordings for some time in many audiophile magazines. Many of the recordings contained in the Binaural Source catalogue are very difficult to obtain otherwise. Highly recommended.

Heyday Records/Algorithm Records
2325 3rd Street, Suite 339
San Francisco, CA 94107
Ron Gompertz has used a HEAD acoustics dummy head to create many interesting studio-based binaural recordings on the Heyday label (e.g., Clubfoot Orchestra, Connie Champagne). Along with writer Lisa Palac, he produces a series of erotic 3-D sound recordings, the "Cyborgasm" series, available from Algorithm Records. Heyday/Algorithm is currently producing a number of books on tape using various 3-D audio technologies.

Patents

An investigation of 3-D sound techniques contained in U.S. patents can be helpful to the advanced reader, and simultaneously misleading for the uninitiated. The foremost thing to remember is that patents are reviewed by legal experts, but not necessarily by technically informed persons. For instance, U.S. patent 4,535,475, "Audio Reproduction Apparatus," filed in 1982 and granted in 1985, describes in careful legal detail "a great advance in the field of portable audio equipment." It is essentially a battery, two speakers, and a radio inside a cabinet: i.e., a "boom box" portable stereo. Nor is there usually any perceptual validation or other proof to the claims that are made within patents beyond citation of secondary sources. The following are but a few of the U.S. patents that are frequently mentioned in connection with 3-D sound. Many patents also have been filed in Japan, Germany, and elsewhere.

U.S. Patent 1,124,580:   **Method and Means for Localizing Sound Reproduction,** by Edward H. Amet (January 1912).
A wonderful description of an early spatial sound playback device.

U.S. Patent 4,731,848: **Spatial Reverberator,** by Gary Kendall and William Martens (March 1988).
The patent is owned by Northwestern University, Evanston, Illinois.

U.S. Patent 4,774,515:   **Attitude Indicator,** by Bo Gehring (September 1988).
Head-tracked audio described for aeronautical applications.

U.S. Patent 4,817,149: **Three-dimensional Auditory Display Apparatus and Method Utilizing Enhanced Bionic Emulation of Human Binaural Sound Localization**, by Peter H. Myers (March 1989).
Notable for its inclusion of Blauert's "boosted bands" (described in Chapter 2) for elevation simulation.

U.S. Patent 5,046,097: **Sound Imaging Process,** by Danny Lowe and John Lees (September 1991).
This patent is the basis of the commercial product known as QSound.

U.S. Patent 5,173,944: **Head-related Transfer Function Pseudo-stereophony,** by Durand R. Begault (December 1992).
This patent was described in Chapter 5 and is licensed to NASA.

References

Ahnert, W., and Feistel, R. (1991). Binaural Auralization from a sound system simulation program. In *91st Audio Engineering Society Convention* (Preprint No. 3127). New York: Audio Engineering Society.

Allen, I. (1991). Matching the Sound to the Picture. (Report No. S91/9146). Dolby Laboratories.

Allen, J. B., and Berkley, D. A. (1979). Image Model for efficiently modeling small–room acoustics. *Journal of the Acoustical Society of America*, **65**, 943–950.

Ando, Y. (1985). *Concert Hall Acoustics*. Berlin: Springer-Verlag.

ANSI (1989). American National Standard: Method for measuring the intelligibility of speech over communication systems. (Report No. S3.2–1989). American National Standards Institute.

Arnold, D. (1959). The significance of cori spezzati. *Music and Letters*, **40**, 5–11.

Asano, F., Suzuki, Y., and Sone, T. (1990). Role of spectral cues in median plane localization. *Journal of the Acoustical Society of America*, **88**, 159–168.

Barron, M. (1971). The subjective effects of first reflections in concert halls — the need for lateral reflections. *Journal of Sound and Vibration*, **15**, 475–494.

Barron, M., and Marshall, A. H. (1981). Spatial impression due to early lateral reflections in concert halls: the derivation of a phsyical measure. *Journal of Sound and Vibration*, **77**, 211–232.

Bassett, I. G., and Eastmond, E. J. (1964). Echolocation: Measurement of Pitch versus Distance for Sounds Reflected from a Flat Surface. *Journal of the Acoustical Society of America*, **36**, 911–916.

Batteau, D. W. (1966). A study of acoustical multipath systems. (Report No. NONR 494-00). United States Navy Office of Naval Research.

Batteau, D. W. (1968). Listening with the naked ear. In S. J. Freedman (Ed.), *The Neuropsychology of spatially oriented behavior*. Homewood, IL: Dorsey Press.

Bauer, B. B. (1961). Stereophonic earphones and binaural loudspeakers. *Journal of the Audio Engineering Society*, **9**, 148–151.

Begault, D. R. (1987). Control of auditory distance. Ph.D. Dissertation, University of California, San Diego.

Begault, D. R. (1991a). Preferred sound intensity increase for sensation of half distance. *Perceptual and Motor Skills*, **72**, 1019–1029.

Begault, D. R. (1991b). Challenges to the successful implementation of 3–D sound. *Journal of the Audio Engineering Society*, **39**, 864–870.

Begault, D. R. (1992a). Perceptual effects of synthetic reverberation on three–dimensional audio systems. *Journal of the Audio Engineering Society*, **40**, 895–904.

Begault, D. R. (1992b). Perceptual similarity of measured and synthetic HRTF filtered speech stimuli. *Journal of the Acoustical Society of America*, **92**, 2334.

Begault, D. R. (1992c). Binaural Auralization and Perceptual Veridicality. In *93rd Audio Engineering Society Convention* (Preprint No. 3421). New York: Audio Engineering Society.

Begault, D. R. (1992d). Head–related transfer function pseudo–stereophony. (United States Patent 5,173,944). Washington, DC: Commisioner of Patents and Trademarks.

Begault, D. R. (1992e). Audio Spatialization Device for Radio Communications. (Patent disclosure ARC 12013–1CU). NASA Ames Research Center.

Begault, D. R. (1993a). Call sign intelligibility improvement using a spatial auditory display: applications to KSC speech communications. In *Proceedings of the Seventh Annual Workshop on Space Operations Applications and Research (SOAR '93)*. Houston: NASA Johnson Space Center.

Begault, D. R. (1993b). Head–up Auditory Displays for Traffic Collision Avoidance System Advisories: A Preliminary Investigation. *Human Factors*, **35**, 707–717.

Begault, D. R., and Erbe, T. R. (1993). Multi–channel spatial auditory display for speech communications. In *95th Audio Engineering Society Convention* (Preprint No. 3707). New York: Audio Engineering Society.

Begault, D. R., and Pittman, M. T. (1994). 3–D Audio Versus Head Down TCAS Displays. (Report No. 177636). NASA Ames Research Center.

Begault, D. R., Stein, N., and Loesche, V. (1991). Advanced Audio Applications in the NASA Ames Advanced Cab Flight Simulator (Unpublished report).

Begault, D. R., and Wenzel, E. M. (1993). Headphone Localization of Speech. *Human Factors*, **35**, 361–376.

Beranek, L. L. (1962). *Music, Acoustics and Architecture*. New York: Wiley.

Beranek, L. L. (1992). Concert Hall Acoustics—1992. *Journal of the Acoustical Society of America*, **92**, 1–39.

Blattner, M., and Dannenberg, R. (Eds.) (1992). *Multimedia Interface Design*. New York: Addison–Wesley.

Blattner, M. M., Smikawa, D. A., and Greenberg, R. M. (1989). Earcons and Icons: their stucture and common desing principles. *Human–Computer Interaction*, **4**, 11–44.

Blauert, J. (1969). Sound localization in the median plane. *Acustica*, **22**, 205–213.

Blauert, J. (1983). *Spatial hearing. The Psychophysics of Human Sound Localization* (J. Allen, Trans.). Cambridge, MA: MIT Press.

Blauert, J., and Cobben, W. (1978). Some consideration of binaural crosscorrelation analysis. *Acustica*, **39**, 96–104.

Blauert, J., and Laws, P. (1973). Verfahren zur orts– und klanggetrauen Simulation von Lautsprecherbeschallungen mit Hilfe von Kopfhörern [Process for positionally and tonally accurate simulation of sound presentation over loudspeakers using headphones]. *Acustica*, **29**, 273–277.

Bloom, P. J. (1977). Creating source elevation illusions by spectral manipulation. *Journal of the Audio Engineering Society*, **61**, 820–828.

Bly, S. (1982). Presenting information in sound. In *Proceedings of the CHI '82 Conference on Human Factors in Computing Systems*. New York: The Association for Computing Machinery.

Borish, J. (1984). Electronic Simulation of Auditorium Acoustics. Ph.D. Dissertation, Stanford University.

Borish, J. (1986). Some New Guidelines for Concert Hall Design Based on Spatial Impression. In *12th International Congress on Acoustics*. Vancouver: Canadian Acoustical Association.

Bregman, A. S. (1990). *Auditory Scene Analysis: The Perceptual Organization of Sound*. Cambridge, MA: MIT Press.

Bronkhorst, A. W., and Plomp, R. (1988). The effect of head–induced interaural time and level differences on speech intelligibility in noise. *Journal of the Acoustical Society of America*, **83**, 1508–1516.

Bronkhorst, A. W., and Plomp, R. (1992). Effect of multiple speechlike maskers on binaural speech recognition in normal and impaired hearing. *Journal of the Acoustical Society of America*, **92**, 3132–3139.

Bryson, S., and Gerald–Yamasaki, M. (1992). The Distributed Virtual Windtunnel. (Report No. RNR–92–010). NASA Ames Research Center.

Buford, J. (Ed.) (1994). *Multimedia systems*. Reading, MA: ACM Press/Addison-Wesley.

Burger, J. F. (1958). Front–back discrimination of the hearing system. *Acustica*, **8**, 301–302.

Burgess, D. A. (1992). Real–time audio spatialization with inexpensive hardware. (Report No. GIT–GVU–92–20). Graphics Visualization and Usability Center, Georgia Institute of Technology.

Burkhard, M. D., and Sachs, R. M. (1975). Anthropometric manikin for acoustic research. *Journal of the Acoustical Society of America*, **58**, 214–222.

Butler, R. A. (1987). An analysis of the monaural displacement of sound in space. *Perception and Psychophysics*, **41**, 1–7.

Butler, R. A., and Belendiuk, K. (1977). Spectral cues utilized in the localization of sound in the median sagittal plane. *Journal of the Acoustical Society of America*, **61**, 1264–1269.

Butler, R. A., Levy, E. T., and Neff, W. D. (1980). Apparent distance of sounds recorded in echoic and anechoic chambers. *Journal of Experimental Psychology*, **6**, 745–750.

Buxton, W., Gaver, W., and Bly, S. (1989). The use of non–speech audio at the interface. (Tutorial No. 10). Unpublished tutorial notes from the CHI 1989 conference, Austin, TX.

Cherry, E. C. (1953). Some experiments on the recognition of speech with one and two ears. *Journal of the Acoustical Society of America*, **25**, 975–979.

Cherry, E. C., and Taylor, W. K. (1954). Some further experiments on the recognition of speech with one and with two ears. *Journal of the Acoustical Society of America*, **26**, 549–554.

Chowning, J. M. (1971). The Simulation of Moving Sound Sources. *Journal of the Audio Engineering Society*, **19**, 2–6.

Cohen, M. (1993). Integrating graphical and audio windows. *Presence: Teleoperators and Virtual Environments*, **1**, 468–481.

Cohen, M., and Koizumi, N. (1991). Audio windows for binaural telecommunication. In *Proceedings of the Joint meeting of Human Communication Committee and Speech Technical Committee* (Paper 91–242). Tokyo: Institute of Electronics, Information and Communication Engineers.

Coleman, P. (1963). An analysis of cues to auditory depth perception in free space. *Psychological Bulletin*, **60**, 302–315.

Coleman, P. D. (1962). Failure to localize the source distance of an unfamiliar sound. *Journal of the Acoustical Society of America*, **34**, 345–346.

Coleman, P. D. (1968). Dual role of frequency spectrum in determination of auditory distance. *Journal of the Acoustical Society of America*, **44**, 631–632.

Cooper, D. H., and Bauck, J. L. (1989). Prospects for transaural recording. *Journal of the Audio Engineering Society*, **37**, 3–19.

Cotzin, M., and Dallenbach, K. M. (1950). "Facial Vision": The role of pitch and loudness in the perception of obstacles by the blind. *American Journal of Psychology*, **63**, 485–515.

CRE (1993). The Acoustic Room Simulator. (Unpublished report). Crystal River Engineering, Inc.

Cremer, L., and Muller, H. A. (1982): *Principles and Applications of Room Acoustics [Die wissenschaftlichen Grundlagen der Raumakustik]* (T. J. Schultz, trans.). London: Applied Science.

Crispien, K., and Petrie, H. (1993). Providing access to GUIs for blind people Using a multimedia system based on spatial audio presentation. In *95th Audio Engineering Society Convention* (Preprint No. 3738). New York: Audio Engineering Society.

D'Antonio, P., Konnert, J., and Kovitz, P. (1993). Disc Project: Auralization using directional scattering coefficients. In *95th Audio Engineering Society Convention* (Preprint No. 3727). New York: Audio Engineering Society.

Dalenbäck, B.–I., Kleiner, M., and Svensson, P. (1993). Audibility of changes in geometric shape, source directivity, and absortpive treatment–experiments in Auralization. *Journal of the Audio Engineering Society*, **41**, 905–913.

Davis, G., and Jones, R. (1989). *Sound Reinforcement Handbook* (2nd Ed.) Milwaukee, WI: Hal Leonard Publishing.

Deutsch, D. (1983). Auditory illusions, handedness, and the spatial environment. *Journal of the Audio Engineering Society*, **31**, 607–623.

Dodge, C., and Jerse, T. A. (1985). *Computer music: Synthesis, Composition, and Performance*. New York: Schirmer Books.

Doll, T. J., Gerth, J. M., Engelman, W. R., and Folds, D. J. (1986). Development of simulated directional audio for cockpit applications. (Report No. AAMRL–TR–86–014). Wright–Patterson Airforce Base, OH: Armstrong Aerospace Medical Research Laboratory.

Durlach, N. I., and Colburn, H. S. (1978). Binaural phenomena. In E. C. Carterette and M. P. Friedman (Eds.), *Handbook of Perception. Volume 4: Hearing*. New York: Academic Press.

Durlach, N. I., Rigopulos, A., Pang, X. D., Woods, W. S., Kulkkarni, A., Colburn, H. S., and Wenzel, E. (1992). On the externalization of auditory images. *Presence*, **1**, 251-257.

Ellis, S. R., Kaiser, M. K., and Grunwald, A. J. (Eds.) (1991). *Pictorial communication in virtual and real environments*. London: Taylor and Francis.

Ericson, M., McKinley, R., Kibbe, M., and Francis, D. (1993). Laboratory and In–flight experiments to evaluate 3–D audio technology. In *Seventh Annual Workshop on Space Operations Applications and Research (SOAR '93)* (Conference Publication 3240). Houston: NASA Johnson Space Center.

Feddersen, W. E., Sandel, T. T., Teas, D. C., and Jeffress, L. A. (1957). Localization of high–frequency tones. *Journal of the Acoustical Society of America*, **29**, 988–991.

Fisher, H., and Freedman, S. J. (1968). The role of the pinnae in auditory localization. *Journal of Auditory Research*, **8**, 15–26.

Fisher, S. S., Coler, C., McGreevy, M. W., and Wenzel, E. M. (1988). Virtual interface environment workstations. In *32nd Annual Meeting of the Human Factors and Ergonomics Society*. Santa Monica, CA: Human Factors and Ergonomics Society.

Fisher, S. S., McGreevy, M., Humphries, J., and Robinett, W. (1986). Virtual Environment Display System. In *ACM 1986 Workshop on Interactive 3D Graphics*. New York: Association for Computing Machinery.

Foley, D. (1987, April). Interfaces for Advanced Computing. *Scientific American*.

Foster, S. H., Wenzel, E. M., and Taylor, R. M. (1991). Real–time synthesis of complex acoustic environments (Summary). In *Proceedings of the ASSP (IEEE) Workshop on Applications of Signal Processing to Audio and Acoustics*. New York: IEEE Press.

Gardner, M. B. (1968). Historical background of the Haas and/or precedence effect. *Journal of the Acoustical Society of America*, **43**, 1243–1248.

Gardner, M. B. (1969). Distance estimation of 0 degree or apparent 0 degree–oriented speech signals in anechoic space. *Journal of the Acoustical Society of America*, **45**, 47–53.

Gardner, M. B. (1973). Some monaural and binaural facets of median plane localization. *Journal of the Acoustical Society of America*, **54**, 1489–1495.

Gardner, M. B., and Gardner, R. S. (1973). Problem of localization in the median plane: Effect of pinnae cavity occlusion. *Journal of the Acoustical Society of America*, **53**, 400–408.

Gardner, W. G. (1992). A realtime multichannel room simulator. *Journal of the Acoustical Society of America*, **92**, 2395.

Gatehouse, R. W. (Ed.) (1982). *Localization of Sound: Theory and Applications*. Groton, CT: Amphora Press.

Gaver, W. (1986). Auditory Icons: Using sound in computer interfaces. *Human–Computer Interaction*, **2**, 167–177.

Gaver, W. W. (1989). The Sonic Finder: An interface that uses auditory icons. *Human–Computer Interaction*, **4**, 67–94.

Gehring, B. (1988). Attitude Indicator. (United States Patent 4,774,515). Washington, DC: Commisioner of Patents and Trademarks.

Genuit, K. (1984). A model for the description of outer–ear transmission characterstics [translation of "Ein Modell zur Beschreibung von Außenohrübertragungseigenschaften"]. Ph.D. Dissertation, Rhenish–Westphalian Technical University, Aachen.

Gierlich, H. W. (1992). The Application of Binaural Technology. *Applied Acoustics*, **36**, 219–243.

Gierlich, H. W., and Genuit, K. (1989). Processing Artifical Head Recordings. *Journal of the Audio Engineering Society*, **37**, 34–39.

Gill, H. S. (1984). Review of Outdoor Sound Propogation. (Unpublished technical report). Wilson, Ihrig, and Associates, Oakland, CA.

Gould, G. (1966, April). The Prospects of Recording. *High Fidelity*.

Grantham, D. W. (1986). Detection and discrimination of simulated motion of auditory targets in the horizontal plane. *Journal of the Acoustical Society of America*, **79**, 1939–49.

Griesinger, D. (1989). Equalization and spatial equalization of dummy head recordings for loudspeaker reproduction. *Journal of the Audio Engineering Society*, **37**, 20–29.

Griesinger, D. (1990). Binaural techniques for music reproduction. In *Proceedings of the AES 8th International Conference*. New York: Audio Engineering Society.

Griesinger, D. (1993). Quantifying Musical Acoustics through Audibility. Unpublished transcript of the Knudsen Memorial Lecture, Denver Acoustical Society.

Haas, H. (1972). The influence of a single echo on the audibility of speech. *Journal of the Audio Engineering Society*, **20**, 146–159.

Hammershøi, D., Møller, H., Sørensen, M., and Larsen, K. (1992). Head–Related Transfer Functions: Measurements on 24 Subjects. In *92nd Audio Engineering Society Convention* (Preprint No. 3289). New York: Audio Engineering Society.

Handel, S. (1989). *Listening: An Introduction to the Perception of Auditory Events*. Cambridge, MA: MIT Press.

Hanson, R. L., and Kock, W. E. (1957). Intereting effect produced by two loudspeakers under free space conditions. *Journal of the Acoustical Society of America*, **29**, 145.

Harris, C. M. (1966). Absorption of sound in air vs. humidity and temperature. *Journal of the Acoustical Society of America*, **40**, 148–159.

Hartmann, W. M. (1983). Localization of sound in rooms. *Journal of the Acoustical Society of America*, **74**, 1380–1391.

Hebrank, J., and Wright, D. (1974). Spectral cues used in the localization of sound sources on the median plane. *Journal of the Acoustial Society of America*, **56**, 1829–1834.

Heinz, R. (1993). Binaural room simulation based on an image source model with addition of statistical methods to include the diffuse sound scattering of walls and to predict the reverberant tail. *Journal of Applied Acoustics*, **38**, 145–160.

Helmholtz, H. (1877). *On the Sensations of Tone As a Psychological Basis for the Theory of Music* (A. Ellis, Trans.; 2nd English edition, 1954). New York: Dover.

Henning, G. B. (1974). Detectability of interaural delay in high—frequency complex waveforms. *Journal of the Acoustical Society of America*, **55**, 84–90.

Holt, R. E., and Thurlow, W. R. (1969). Subject orientation and judgement of distance of a sound sourcee. *Journal of the Acoustical Society of America*, **46**, 1584–5.

ISO (1993). Information technology—Coding of moving pictures and associated audio for digital storage media at up to about 1.5 Mbit/s. Part 3: Audio. (Report No. ISO/IEC 11172–3). International Organization for Standardization.

Jacob, K. D. (1989). Correlation of speech intelligibility tests in reverberant rooms with three predictive algorithms. *Journal of the Audio Engineering Society*, **37**, 1020–1030.

Jacob, K. D., Jørgensen, M., and Ickler, C. B. (1992). Verifying the accuracy of audible simulation (auralization) systems. *Journal of the Acoustical Society of America*, **92**, 2395.

Jacobson, L. (1991, October). Psychoacoustic Satisfaction: Exploring the Frontiers of 3–D Sound. *Mix*.

Jacobson, L. (Ed.) (1992). *Cyberarts. Exploring Art and Technology*. San Francisco: Miller Freeman.

Jeffress, L. (1948). A place theory of sound localization. *Journal of comparative and physiological psychology*, **61**, 468–486.

Jørgensen, M., Ickler, C. B., and Jacob, K. D. (1993). Using subject-based testing to evaluate the accuracy of an audible simulation system. *95th Audio Engineering Society Convention* (Preprint No. 3725). New York: Audio Engineering Society.

Junqua, J. C. (1993). The Lombard reflex and its role on human listeners and automatic speech recognizers. *Journal of the Acoustical Society of America*, **93**, 510–524.

Kalawsky, R. S. (1993). *The Science of Virtual Reality and Virtual Environments*. Wokingham, England: Addison–Wesley.

Kellogg, W. N. (1962). Sonar system of the blind. *Science*, **137**, 399–404.

Kendall, G., Martens, W. L., and Wilde, M. D. (1990). A spatial sound processor for loudspeaker and headphone reproduction. In *Proceedings of the AES 8th International Conference*. New York: Audio Engineering Society.

Kendall, G. S., and Martens, W. L. (1984). Simulating the cues of spatial hearing in natural environments. In *Proceedings of the 1984 International Computer Music Conference*. San Francisco: International Computer Music Association.

Kendall, G. S., and Rodgers, C. A. P. (1982). The simulation of three–dimensional localization cues for headphone listening. In *Proceedings of the 1982 International Computer Music Confernece*. San Francisco: International Computer Music Association.

Kistler, D. J. (1992). A model of head–related transfer functions based on principal components analysis and minimum–phase reconstruction. *Journal of the Acoustical Society of America*, **91**, 1637–1647.

Kleiner, M., Dalenbäck, B. I., and Svensson, P. (1993). Auralization—an overview. *Journal of the Audio Engineering Society*, **41**, 861–875.

Knudsen, E. I., and Brainard, M. S. (1991). Visual Instruction of the Neural Map of Auditory Space in the Developing Optic Tectum. *Science*, **253**, 85–87.

Knudsen, E. I., and Konishi, M. (1978). A neural map of auditory space in the owl. *Science*, **200**, 795–797.

Knudsen, V. O. (1963, November). Architectural Acoustics. *Scientific American*.

Korenaga, Y., and Ando, Y. (1993). A sound–field simulation system and its application to a seat–selection system. *Journal of the Audio Engineering Society*, **41**, 920–930.

Krueger, M. W. (1991). *Artifical Reality II*. Reading, MA: Addison–Wesley.

Kryter, K. D. (1972). Speech Communication. In H. P. V. Cott and R. G. Kinkade (Eds.), *Human Engineering Guide to Equipment Design*. Washington, DC: McGraw–Hill.

Kuttruff, K. H. (1993). Auralization of impulse responses modeled on the basis of ray–tracing results. *Journal of the Audio Engineering Society*, **41**, 876–880.

Kyma (1991). KYMA advertising literature. Symbolic Sound Corporation, Champaign, IL.

Laurel, B. (1993). *Computers as Theatre*. Reading, MA: Addison–Wesley.

Laurie, M. (1980). Echo music. In S. Sadie (Ed.), *The New Grove Dictionary of Music and Musicians*. London: MacMillan Publishers.

Laws, P. (1973). Auditory distance perception and the problem of "in–head localization" of sound images [Translation of "Entfernungshören und das Problem der Im–Kopf–Lokalisiertheit von Hörereignissen,"*Acustica*, **29**, 243–259]. NASA Technical Translation TT–20833.

Leakey, D. M., Sayers, B. M., and Cherry, C. (1958). Binaural fusion of low and high frequency sounds. *Journal of the Acoustical Society of America*, **30**, 322.

Lehnert, H., and Blauert, J. (1992). Principals of Binaural Room Simulation. *Applied Acoustics*, **36**, 259–91.

Levitt, H., and Rabiner, L. R. (1967). Binaural release from masking for speech and gain in intelligibility. *Journal of the Acoustcal Society of America*, **42**, 601–608.

Lowe, D., and Lees, J. (1991). Sound Imaging Process. (United States Patent 5,046,097). Washington, DC: Commisioner of Patents and Trademarks.

Ludwig, L. F., Pincever, N., and Cohen, M. (1990). Extending the notion of a window system to audio. *Computer*, **23**, 66–72.

Makous, J. C., and Middlebrooks, J. C. (1990). Two–dimensional sound localization by human listeners. *Journal of the Acoustical Society of America*, **87**, 2188–2200.

Martens, W. L. (1987). Principal components analysis and resynthesis of spectral cues to perceived direction. In *Proceedings of the 1987 International Computer Music Conference*. San Francisco: International Computer Music Association.

Martens, W. L. (1991). Directional Hearing on the Frontal Plane: Necessary and Sufficient Spectral Cues. Ph.D. Dissertation, Northwestern University.

Mason, W. (1969). The architecture of St. Mark's Cathedral and the Venitian polychoral style. In J. Pruett (Ed.), *Studies in Musicology*. Chapel Hill, NC: University of North Carolina Press.

McClellan, J. H., Parks, T. W., and Rabiner, L. R. (1979). FIR Linear Phase Filter Design Program. In *Programs for Digital Signal Processing*. New York: IEEE Press.

McGregor, P., Horn, A. G., and Todd, M. A. (1985). Are familiar sounds ranged more accurately? *Perceptual and Motor Skills*, **61**, 1082.

McKinley, R. L., Ericson, M. A., and D'Angelo, W. R. (1994). 3-D Auditory Displays: Development, Applications and Performance. *Aviation Space and Environmental Medicine*, **65**, a31-a38.

Mehrgardt, S., and Mellert, V. (1977). Transformation characteristics of the external human ear. *Journal of the Acoustical Society of America*, **61**, 1567–1576.

Mershon, D. H., and Bowers, J. N. (1979). Absolute and relative distance cues for the auditory perception of egocentric distance. *Perception*, **8**, 311–322.

Mershon, D. H., and King, L. E. (1975). Intensity and reverberation as factors in the auditory perception of egocentric distance. *Perception and Psychophysics*, **18**, 409–415.

Meyer, K., and Applewhite, H. L. (1992). A survey of position trackers. *Presence*, **1**, 173–200.

Mezrich, J. J., Frysinger, S., and Slivjanovski, R. (1984). Dynamic representation of multivarite time series data. *Journal of the American Statistical Association*, **79**, 34–40.

Middlebrooks, J. C. (1992). Narrow–band sound localization related to external ear acoustics. *Journal of the Acoustical Society of America*, **92**, 2607–2624.

Middlebrooks, J. C., and Green, D. M. (1991). Sound Localization by Human Listeners. *Annual Review of Psychology*, **42**, 135–159.

Middlebrooks, J. C., Makous, J. C., and Green, D. M. (1989). Directional sensitivity of sound–pressure levels in the human ear canal. *Journal of the Acoustical Society of America*, **86**, 89–108.

Mills, W. (1972). Auditory localization. In J. V. Tobias (Ed.), *Foundations of Modern Auditory Theory*. New York: Academic Press.

Mochimaru, A. (1993). A study of the practicality and accuracy of impulse response calculations for the auralization of sound system design. *Journal of the Audio Engineering Society*, **41**, 881–893.

Møller, H. (1992). Fundamentals of Binaural Technology. *Applied Acoustics*, **36**, 171–218.

Moore, B. C. J. (1989). *Introduction to the Psychology of Hearing* (3rd Ed.) New York: Academic Press.

Moore, B. C. J., Oldfield, S. R., and Dooley, G. J. (1989). Detection and discrimination of spectral peaks and notches at 1 and 8 kHz. *Journal of the Acoustical Society of America*, **85**, 820–836.

Moore, F. R. (1983). A general model for the spatial processing of sounds. *Computer Music Journal*, **7**, 6–15.

Moore, F. R. (1986). Personal communication.

Moore, F. R. (1987). The Dysfunctions of MIDI. In *Proceedings of the 1987 International Computer Music Conference*. San Francisco: International Computer Music Association.

Moore, F. R. (1990). *Elements of Computer Music*. Englewood Cliffs, NJ: Prentice-Hall.

Moorer, J. A. (1985). About This Reverberation Business. In C. Roads and J. Strawn (Eds.), *Foundations of Computer Music*. Cambridge, MA: MIT Press.

Morimoto, M., and Ando, Y. (1982). On the simulation of sound localiztion. In R. W. Gatehouse (Ed.), *Localization of Sound: Theory and Applications*. Groton, CT: Amphora Press.

Mumma, G. (1986). Audioearotica. *In* Center for Music Experiment and Related Research Annual Report 1986–7. La Jolla, CA: University of California, San Diego.

Musicant, A. D., and Butler, R. A. (1984). The influence of pinnae–based spectral cues on sound localization. *Journal of the Acoustical Society of America*, **75**, 1195–1200.

Myers, P. H. (1989). Three–dimensional Auditory Display Apparatus and Method Utilizing Enhanced Bionic Emulation of Human Binaural Sound Localization. (United States Patent 4,817,149). Washington, DC: Commisioner of Patents and Trademarks.

Neumann (1992). Operating Instructions for the KU 100 Dummy Head. (Report No. 12231 80101). Georg Neumann GmbH, Berlin.

Noble, W. (1987). Auditory localization in the vertical plane: Accuracy and constraint on bodily movement. *Journal of the Acoustical Society of America*, **82**, 1631–1636.

Oldfield, S. R., and Parker, S. P. A. (1984a). Acuity of sound localisation: a topography of auditory space. I. Normal hearing conditions. *Perception*, **13**, 581–600.

Oldfield, S. R., and Parker, S. P. A. (1984b). Acuity of sound localisation: a topography of auditory space. II. Pinnae cues absent. *Perception*, **13**, 601–617.

Oldfield, S. R., and Parker, S. P. A. (1986). Acuity of sound localisation: a topography of auditory space. III. Pinnae cues absent. *Perception*, **15**, 67–81.

Oppenheim, A. V., and Schafer, R. W. (1975). *Digital Signal Processing*. Englewood Cliffs, NJ : Prentice-Hall.

Oppenheim, A. V., and Schafer, R. W. (1989). *Discrete–time Signal Processing*. Englewood Cliffs, NJ : Prentice-Hall.

Palmer, E., and Degani, A. (1991). Electronic Checklists: Evaluation of Two Levels of Automation. In *Proceedings of the Sixth Symposium on Aviation Psychology*. Columbus, OH: Ohio State University.

Patterson, R. (1982). Guidelines for auditory warning systems on civil aircraft. (Report No. 82017). London, UK: Civil Aviation Authority.

Perrott, D. R., Sadralodabai, T., Saberi, K., and Strybel, T. Z. (1991). Aurally aided visual search in the central visual field: effects of visual load and visual enhancement of the target. *Human Factors*, **33**, 389–400.

Persterer, A. (1991). Binaural simulation of an "Ideal control room for headphone reproduction. In *90th Audio Engineering Society Convention* (Preprint No. 3062). New York: Audio Engineering Society.

Pierce, J. R. (1983). *The Science of Musical Sound*. New York: Scientific American Books.

Pimentel, K., and Teixeira, K. (1993). *Virtual Reality. Through the New Looking Glass*. Blue Ridge Summit, PA: Windcrest.

Plenge, G. (1974). On the difference between localization and lateralization. *Journal of the Acoustical Society of America*, **56**, 944–951.

Pohlmann, K. C. (1992). *The Compact Disc Handbook*. Madison, WI: A–R Editions.

Pollack, I., and Pickett, J. M. (1958). Stereophonic listening and speech intelligbility. *Journal of the Acoustical Society of America*, **30**, 131–133.

Rabinovich, A. V. (1936). The effect of distance in the broadcasting studio. *Journal of the Acoustcial Society of America*, **7**, 199–203.

Rayleigh, L. (1907). On our perception of sound direction. *Philosophical magazine*, **13**, 214–232.

Rheingold, H. (1991). *Virtual Reality*. New York: Summit Books.

Rice, C. E. (1967). Human echo perception. *Science*, **155**, 399–404.

Rice, C. E., Feinstein, S. H., and Schusterman, R. J. (1965). Echo–detection ability of the blind: Size and distance factors. *Journal of Experimental Psychology*, **70**, 246–251.

Rife, D. D., and Vanderkooy, J. (1989). Transfer–function measurements with maximum–length sequences. *Journal of the Audio Engineering Society*, **37**, 419–444.

Rodgers, C. A. P. (1981). Pinna transformations and sound reproduction. *Journal of the Audio Engineering Society*, **29**, 226–234.

Roederer, J. G. (1975). *Introduction to the Physics and Psychophysics of Music.* (2nd Ed.). New York: Springer-Verlag.

Roffler, S. K., and Butler, R. A. (1968). Factors that influence the localization of sound in the vertical plane. *Journal of the Acoustical Society of America,* **43**, 1255–1259.

Runstein, R. E., and Huber, D. M. (1989). *Modern Recording Techniques* (3rd Ed.). Indianapolis, IN: Howard W. Sams.

Sakamoto, N., Gotoh, T., and Kimura, Y. (1976). On "out–of–head localization" in headphone listening. *Journal of the Audio Engineering Society,* **24**, 710–716.

Sayers, B. (1964). Acoustic–image lateralization judgement with binaural tones. *Journal of the Acoustical Society of America,* **36**, 923–926.

Scaletti, C., and Craig, A. B. (1991). Using sound to extract meaning from complex data. In *SPIE Extracting Meaning from Complex Data: Processing, Display, Interaction II.* (Volume No. 1459-V). San Jose: SPIE—The International Society for Optical Engineerng.

Scharf, B. (1970). Critical bands. In J. V. Tobias (Ed.), *Foundations of Modern Auditory Theory.* New York: Academic Press.

Schouten, J. F. (1968). The Perception of Timbre. In *Reports of the 6th International Congress on Acoustics* (Paper GP–6). Tokyo: Maruzen.

Schroeder, M. R. (1970). Digital simulation of sound transmission in reverberant spaces. *Journal of the Acoustical Society of America,* **47**, 424–431.

Schroeder, M. R. (1973). Computer models for concert hall acoustics. *American Journal of Physics,* **41**, 461–471.

Schroeder, M. R. (1980, October). Toward better acoustics for concert halls. *Physics Today.*

Schroeder, M. R. (1984). Progress in architectural acoustics and artificial reverberation: Concert hall acoustics and number theory. *Journal of the Audio Engineering Society,* **32**, 194–203.

Schroeder, M. R., and Atal, B. S. (1963). Computer simulation of sound transmissioon in rooms. In *IEEE International Convention Record* (7). New York: IEEE Press.

Schroeder, M. R., and Logan, B. F. (1961). "Colorless" artificial reverberation. *Journal of the Audio Engineering Society*, **9**, 192–197.

Searle, C. L., Braida, L. D., Cuddy, D. R., and Davis, M. F. (1975). Binaural pinnae disparity: another auditory localization cue. *Journal of the Acoustical Society of America*, **57**, 448–455.

Searle, C. L., Braida, L. D., Davis, M. F., and Colburn, H. S. (1976). Model for auditory localization. *Journal of the Acoustical Society of America*, **60**, 1164–1175.

Shaw, E. A., and Teranishi, R. (1968). Sound pressure generated in an external–ear replica and real human ears by a nearby sound source. *Journal of the Acoustical Society of America*, **44**, 240–249.

Shaw, E. A. G. (1974). Transformation of sound pressure level from the free field to the eardrum in the horizontal plance. *Journal of the Acoustical Society of America*, **56**, 1848–1861.

Shaw, E. A. G., and Vaillancourt, M. M. (1985). Transformation of sound–pressure level from the free field to the eardrum presented in numerical form. *Journal of the Acoustical Society of America*, **78**, 1120–1122.

Sheeline, C. W. (1982). An investigation of the effects of direct and reverberant signal interaction on auditory distance perception. Ph.D. Dissertation, Stanford University.

Shinn–Cunningham, B., Delhorne, L., Durlach, N., and Held, R. (1994). Adaptation to supernormal auditory localization cues as a function of rearrangement strength. *Journal of the Acoustical Society of America*, **95**, 2896.

Smith, J. O. (1985). Introduction to Digital Filter Theory. (Report No. STAN–M–20). Center for Computer Research in Music and Acoustics, Stanford University.

Smith, S., Bergeron, R. D., and Grinstein, G. G. (1990). Stereophonic and surface sound generation for exploratory data. In *CHI 90 ACM Conference on Computer–Human Interaction*. New York: The Association for Computing Machinery.

Sorkin, R. D., Wightman, F. L., Kistler, D. S., and Elvers, G. C. (1991). An exploratory study of the use of movement–correlated cues in an auditory head–up display. *Human Factors*, **31**, 161–166.

Steickart, H. W. (1988). Head–related Stereophony: New Knowledge Gained in Productions and Reception. (Transcription of lecture at the 15th Tonmeistertagung, Mainz, Germany). In *The Dummy Head—Theory and Practice*. Berlin: Georg Neumann GmbH.

Stevens, S. S. (1955). The measurement of loudness. *Journal of the Acoustical Society of America*, **27**, 815–829.

Stevens, S. S., and Guirao, M. (1962). Loudness, reciprocality, and partition scales. *Journal of the Acoustical Society of America*, **34**, 1466–1471.

Stevens, S. S., and Newman, E. B. (1936). The localization of actual sources of sound. *American Journal of Psychology*, **48**, 297–306.

Strawn, J. (Ed.) (1985). *Digital Audio Signal Processing: An Anthology*. Madison, WI: A-R Editions.

Strybel, T. Z., Manligas, C. L., and Perrott, D. R. (1992). Minimum Audible Movement Angle as a Function of the Azimuth and Elevation of the Source. *Human Factors*, **34**, 267–275.

Theile, G. (1983, February). Study on the standardisation of studio headphones. *EBU Review*.

Theile, G. (1986). On the standardization of the frequency response of high–quality studio headphones. *Journal of the Audio Engineering Society*, **34**, 953–969.

Thielcke, G. A. (1976). *Bird Sounds*. Ann Arbor, MI: University of Michigan Press.

Thurlow, W. R., Mangels, J. W., and Runge, P. S. (1967). Head movements during sound localization. *Journal of the Acoustical Society of America*, **42**, 489–493.

Thurlow, W. R., and Runge, P. S. (1967). Effects of induced head movements on localization of direct sound. *Journal of the Acoustical Society of America*, **42**, 480–487.

Thurstone, L. L. (1927). A Law of Comparative Judgement. *Psychological Review*, **34**, 273–289.

Tierney, J. (1993, September 16). Jung in Motion, Virtually, and Other Computer Fuzz. *The New York Times*, B1–4.

Toole, F. E. (1970). In–head localization of acoustic images. *Journal of the Acoustical Society of America*, **48**, 943–949.

Toole, F. E., and Sayers, B. (1965). Lateralization judgements and the nature of binaural acoustic images. *Journal of the Acoustical Society of America*, **37**, 319–324.

Trahiotis, C., and Bernstein, L. R. (1986). Lateralization of bands of noise and sinusoidally amplitude–modulated tones: effects of spectral locus and bandwidth. *Journal of the Acoustical Society of America*, **79**, 1950–1957.

Van, J. (1994, February 8). Sounding the alarm: Noisy medical alert systems are getting out of hand. *San Jose Mercury News*, 1E–2E.

von Békésy, G. (1960). *Experiments in Hearing* (E. G. Wever, Trans.). New York: McGraw–Hill.

von Békésy, G. (1967). *Sensory Inhibition*. Princetion, NJ: Princeton University Press.

Wallach, H. (1940). The role of head movements and vestibular and visual cues in sound localization. *Journal of Experimental Psychology*, **27**, 339–368.

Wallach, H., Newman, E. B., and Rosenzweig, M. R. (1973). The precedence effect in sound localization. *Journal of the Audio Engineering Society*, **21**, 817–826.

Warusfel, O., Kahle, E., and Jullien, J. P. (1993). Relationships between objective measurements and perceptual interpretation: The need for considering spatial emission of sound sources. *Journal of the Acoustical Society of America*, **93**, 2281–2282.

Watkins, A. J. (1978). Psychoacoustical aspects of synthesized vertical locale cues. *Journal of the Acoustical Society of America*, **63**, 1152–1165.

Weinrich, S. (1982). The problem of front–back localization in binaural hearing. In Tenth Danavox Symposium on Binaural Effects in Normal and Impaired Hearing. *Scandanavian Audiology*, **11**, Suppl. 15).

Wenzel, E. M., Arruda, M., Kistler, D. J., and Wightman, F. L. (1993). Localization using non–individualized head–related transfer functions. *Journal of the Acoustical Society of America*, **94**, 111–123.

Wenzel, E. M., Fisher, S., Stone, P. K., and Foster, S. H. (1990). A system for three–dimensional acoustic "visualization" in a virtual environment workstation. In *Proceedings of Visualization '90 Conference*. New York: IEEE Press.

Wenzel, E. M., and Foster, S. H. (1993). Perceptual consequences of interpolating head–related transfer functions during spatial synthesis. In *Proceedings of the ASSP (IEEE) Workshop on Applications of Signal Processing to Audio and Acoustics*. New York: IEEE Press.

Wenzel, E. M., Wightman, F. L., and Foster, S. H. (1988). A virtual display system for conveying three–dimensional acoustic information. In *32nd Annual Meeting of the Human Factors and Ergonomics Society*. Santa Monica: Human Factors and Ergonomics Society.

Wenzel, E. M., Wightman, F. L., Kistler, D. J., and Foster, S. H. (1988). Acoustic origins of individual differences in sound localization behavior. *Journal of the Acoustical Society of America*, **84**, S79.

Wexelblat, A. (Ed.) (1993). *Virtual Reality. Applications and Explorations*. Cambridge, MA: Academic Press.

Wightman, F. L., and Kistler, D. J. (1989a). Headphone simulation of free–field listening. I: Stimulus synthesis. *Journal of the Acoustical Society of America*, **85**, 858–867.

Wightman, F. L., and Kistler, D. J. (1989b). Headphone simulation of free–field listening. II: Psychophysical validation. *Journal of the Acoustical Society of America*, **85**, 868–878.

Wightman, F. L., and Kistler, D. J. (1992). The dominant role of low–frequency interaural time differences in sound localization. *Journal of the Acoustical Society of America*, **91**, 1648–1661.

Wightman, F. L., Kistler, D. J., and Arruda, M. (1992). Perceptual consequences of engineering compromises in synthesis of virtual auditory objects. *Journal of the Acoustical Society of America*, **92**, 2332.

Wright, D., Hebrank, J. H., and Wilson, B. (1974). Pinna reflections as cues for localization. *Journal of the Acoustical Society of America*, **56**, 957–962.

Zhou, B., Green, D. M., and Middlebrooks, J. C. (1992). Characterization of external ear impulse responses using Golay codes. *Journal of the Acoustical Society of America*, **92**, 1169–1171.

Zurek, P. M. (1993). Binaural Advantages and Directional Effects in Speech Intelligibility. In G. A. Studebaker and I. Hochberg (Eds.), *Acoustical Factors Affecting Hearing Aid Performance*. Needham Heights, MA.: Allyn and Bacon.

Zwicker, E. (1961). Subdivision of the Audible Frequency Range into Critical Bands (Frequenzgruppen). *Journal of the Acoustical Society of America*, **33**, 248.

Zwicker, E., and Scharf, B. (1965). A model of loudness summation. *Psychological Review*, **72**, 3–26.

Zwislocki, J. (1962). Analysis of the middle–ear function. *Journal of the Acoustical Society of America*, **34**, 1514–1532.

Index